T0243269

Piping Hot Bees and Boisterous Buzz-Runners

Piping Hot Bees and Boisterous Buzz-Runners

20 Mysteries of Honey Bee Behavior Solved

THOMAS D. SEELEY

WITH ILLUSTRATIONS BY MARGARET C. NELSON

PRINCETON UNIVERSITY PRESS
PRINCETON AND OXFORD

Published by Princeton University Press
41 William Street, Princeton, New Jersey 08540
99 Banbury Road, Oxford OX2 6JX

press.princeton.edu

Library of Congress Cataloging-in-Publication Data

Names: Seeley, Thomas D., author. | Nelson, Margaret C., illustrator.
Title: Piping hot bees and boisterous buzz-runners : 20 mysteries of honey bee behavior solved / Thomas D. Seeley ; with illustrations by Margaret C. Nelson
Other titles: 20 mysteries of honey bee behavior solved
Description: Princeton, N.J. ; Oxford : Princeton University Press, [2024] | Includes bibliographical references and index.
Identifiers: LCCN 2023021515 (print) | LCCN 2023021516 (ebook) | ISBN 9780691237695 (hardback) | ISBN 9780691237701 (ebook)
Subjects: LCSH: Bees—Behavior. | Bee culture. | Beehives. | BISAC: NATURE / Animals / Insects & Spiders | SCIENCE / Life Sciences / Biology
Classification: LCC QL569.4 .S44 2024 (print) | LCC QL569.4 (ebook) | DDC 595.79/915—dc23/eng/20230516
LC record available at https://lccn.loc.gov/2023021515
LC ebook record available at https://lccn.loc.gov/2023021516

British Library Cataloging-in-Publication Data is available

Editorial: Alison Kalett and Hallie Schaeffer
Production Editorial: Nathan Carr
Text Design: Carmina Alvarez-Gaffin
Jacket/Cover Design: Karl Spurzem
Production: Danielle Amatucci
Publicity: Matthew Taylor and Kate Farquhar-Thomson
Copyeditor: Annie Gottlieb

Jacket/Cover Credit: peter_waters / Adobe Stock

This book has been composed in Sabon LT Std

Printed on acid-free paper. ∞

Printed in China

1 3 5 7 9 10 8 6 4 2

Dedicated to my family, Robin, Saren, Maira, and Mo,
and to my mentors, Roger A. Morse, Bert Hölldobler,
Edward O. Wilson, Martin Lindauer, and Bernd Heinrich

Contents

Preface

I am a biologist, and for more than 50 years I have studied an insect that I find endlessly fascinating: the honey bee, *Apis mellifera*. What fascinates me most about this bee is its behavior. My goal has been to understand how the 25,000 or so workers in a honey bee colony cooperate to accomplish such feats as finding and occupying a snug nest cavity, furnishing it with beeswax combs, filling these combs with brood and food, and keeping everyone well nourished, comfortably warm, and safe from intruders. Every investigation that I have undertaken was time-consuming, many were physically demanding, and a few were frustrating. Nevertheless, because the behavior of worker honey bees contains so many compelling mysteries, and because I have found it such a joy to solve some of them, I have kept my research efforts focused on studying the habits of these wondrous little creatures.

I have set myself two goals in writing this book. First, I want to share with a general audience—especially fellow beekeepers—the discoveries that my colleagues and I have made about the behavior of honey bees over the last five decades. Second, I want to share what it was like to make these discoveries: the personal experiences that drew us to our studies, the sights and sounds of our detective work, the negative results that proved useful, and the feelings of delight that came with solving mysteries about these bees. In short, my goal is to share both the products and the processes of our investigations.

This book has 20 chapters. Each one addresses a mystery about how the workers in a honey bee colony act and interact within their intricately organized nests. For example, how can thousands of bees live together inside a snug cavity (tree hollow or bee hive) without suffering asphyxia?

How is the queen's chemical signal of her presence transmitted to the thousands of nurse bees in her colony? And why do nectar foragers sometimes produce a strange, jerky behavior called the tremble dance when they return home? The chapters are ordered chronologically, partly to show the course of my studies, but mainly to help make everything clear. Sometimes, the work that I describe in one chapter builds on findings that I have reported in a previous chapter or two.

Each chapter tells a story of scientific discovery. So, besides reporting *what* was learned, I describe *how* my colleagues and I encountered, and then solved, a mystery about the behavior of honey bees. We will see that close observations, careful experiments, enthusiastic teamwork, and often lucky accidents were the key ingredients of these investigations. We will see, too, what general insights about honey bee behavior and social life emerged from each study. And, looking to the future, we will see that the scientific terrain of honey bee behavior still has countless uncharted regions that beckon for exploration.

Our knowledge of the behavior of honey bees has been acquired through the labors of hundreds of people, many of whom have made the study of the species *Apis mellifera* their life's work. In writing this book, I have felt a great debt of gratitude to the men and women whose hard work helps us to appreciate the richness of the lives of these bees. It is impossible for me to express my thanks to all these prior investigators by name, but those whose studies are featured in this book deserve special mention: M. Delia Allen, Colin G. Butler, Harald Esch, Karl von Frisch, Eleonore Hammann, Engel H. Hazelhoff, Martin Lindauer, Waltraud Meyer, Keith N. Slessor, Hayo H. W. Velthuis, Christina Verheijen-Voogd, and Mark L. Winston.

I also feel deeply grateful to the individuals with whom I have worked directly. In alphabetical order, my co-investigators are Pongthep Akratanakul, Madeleine Beekman, Koos Biesmeijer, Susannah Buhrman-Deever, Brigitte Bujok, Nicholas W. Calderone, Scott Camazine, Fred C. Dyer, Robert L. Fathke, Kathryn E. Gardner, Sean R. Griffin,

Barrett A. Klein, Marco Kleinhenz, Susanne Kühnholz, Benjamin B. Land, Roger A. Morse, Jun Nakamura, James C. Nieh, Madeleine M. Ostwald, Kevin M. Passino, David T. Peck, Robin Radcliffe, Juliana Rangel, Clare C. Rittschof, Kevin Schultz, Robin Hadlock Seeley, Michael L. Smith, James Sneyd, Jürgen Tautz, Craig A. Tovey, William F. Towne, P. Kirk Visscher, and Anja Weidenmüller.

Over the years, many undergraduate students at Yale University, Cornell University, the University of Würzburg (Germany), and the University of Zürich (Switzerland), have played critical roles in the field experiments that lie at the heart of this book. These students were critically important because the experiments required collecting data simultaneously at several locations: a swarm and multiple nest boxes on Appledore Island in Maine, or an observation hive and multiple feeding stations at Cranberry Lake in New York. So, I thank heartily all these summer helpers. In temporal succession they are Andrew Swartz and Roy Levien (1985); Mary Eickwort and Oliver Habicht (1987); Samantha Sonnak and Scott Kelley (1989); Kim Bostwick, Stephen Bryant, and James C. Nieh (1990); Erica van Etten and Timothy Judd (1991); Cornelia König, Timothy Judd, and Barrett Klein (1992 and 1993); Susanne Kühnholz and Anja Weidenmüller (1994); Susannah Buhrman (1997 and 1998); Ethan Wolfson-Seeley (2000); Siobhan Cully and Ben Land (2002); Adrian Reich (2003 and 2004); Robert Fathke (2004); Arielle Zimmerman (2006); Marielle Newsome (2007); Sean R. Griffin (2008–2010); Carter Loftus (2012); and Madeleine M. Ostwald (2013).

I am also extremely thankful to the National Science Foundation, the U.S. Department of Agriculture, and the National Geographic Society; and to Katherine Collins, Founder of the Honeybee Capital Foundation, for providing the financial support of many of the studies discussed in this book.

The work described here could not have been done without access to three special study sites in Maine, Connecticut, and New York State. So I express special thanks to the series of directors of the Shoals Marine Laboratory—Professor John M. Kingsbury, Dr. John B. Heiser, and

Professor James R. Morin—who welcomed me to bring hives of honey bees and teams of co-workers out to Appledore Island. This nearly treeless island, which sits about 7 miles (11 kilometers) off the southern coast of Maine, was essential for most of the experimental studies of how a honey bee swarm chooses and then moves to a new homesite. I feel profoundly grateful, too, to Professors David M. Smith (Yale) and Aaron N. Moen (Cornell). They welcomed me to place hives of bees and perform experiments in the Yale-Myers Forest (Connecticut) and the Arnot Forest (New York State). By working in these forests, we have learned much about how a honey bee colony functions in its natural, forest environment. I am forever indebted, too, to the directors of the Cranberry Lake Biological Station: Professors Rainer H. Brocke, William M. Shields, Robin Wall Kimmerer, and Alex Weir (State University of New York College of Environmental Science and Forestry). Over a span of 35 years, they welcomed me and my co-workers to their field biologist's paradise and they allowed us to set up our observation hive, sugar-water feeders, and other apparatus wherever was best for whatever experiment was at hand. I feel a special debt of gratitude to Larry Rathman, Station Manager, for making available his workshop and boats, and for his help in protecting my bee hives from black bears.

I also owe a huge debt of gratitude to three friends who are beekeepers: Ann Chilcott in the Scottish Highlands, Leo Sharashkin in Missouri, and Richard Woodham in Alabama. Their constructive feedback on every chapter has helped me to tell my stories clearly.

Throughout my work on this book, I have been deeply conscious of the mounting debt of gratitude I owe to Margaret C. Nelson, who has cheerfully worked hard to create nearly all the illustrations for this book. Margy's talent for depicting the bees' structures and behaviors, and showing graphically the results of our experiments, has lightened immensely my task of describing in words what we have learned about the behavior of honey bees. I feel deeply grateful, too, to my friend Barrett A. Klein for the drawing of my beloved observation hive inside its special hut and for the depiction of a worker bee engaged in gobbeting. These drawings appear in Chapters 17 and 19.

The library resources of Cornell University have been placed freely at my disposal, and I wish to express special thanks to the staff of the Cornell Library Annex, which holds all the scientific journals in which Karl von Frisch, Martin Lindauer, Colin G. Butler, and other previous investigators published their work. Reading their original reports, many in German, has deepened my sense of connection to my predecessors and my understanding of their studies.

Lastly, I wish to express deep thanks to several other individuals who, like Margy Nelson, have helped me for decades. First is a group of five biologists: Roger A. Morse (Doc), Bert Hölldobler, Edward O. Wilson, Martin Lindauer, and Bernd Heinrich. Over the past 50 years, I have become a better scientist and writer because of their friendship and guidance. Second is Alison Kalett, the Editorial Director for Science (Biology) at Princeton University Press. We met at a conference at Cornell in 2008, and ever since then Alison has provided steady support and sound advice. Third is my family, which includes my wife, Robin Hadlock Seeley, a marine conservation biologist, who shares my passion for investigating the natural world, and our two daughters, Saren and Maira, who when young were patient while I did my field studies, and as adults have provided lovely encouragement.

Thomas D. Seeley

Ithaca, New York, September 2022

Piping Hot Bees and Boisterous Buzz-Runners

CHAPTER 1

Avoiding Asphyxia

Every journey starts somewhere and sometime. My scientific journey began in my boyhood home, which is a little valley called Ellis Hollow. It lies a few miles east of the small city of Ithaca, in New York State. The time was early June 1963, and I was not quite 11 years old. There were not many houses in Ellis Hollow back then, so not many people used the road that runs through this valley. I enjoyed walking along it slowly, looking and listening. The soft songs of hermit thrushes, veeries, and other forest birds floated out of the woods that sloped up Snyder Hill to the south. The bubbling chatters of bobolinks shot from the unused hayfields that tilted down toward Cascadilla Creek to the north.

One morning, as I approached the massive black walnut tree (*Juglans nigra*) that stands beside Ellis Hollow Road near my parents' house, I heard something strange: a steady, buzzy sort of hum coming from overhead. I looked up and saw thousands of insects flying every which way among the walnut tree's widely spread limbs. Cool! Even cooler was what I noticed next: hundreds of honey bees were landing on this tree's lowest limb, covering an area about the size of a cafeteria tray. Sunlight glinting off their wings had drawn my eyes to their landing zone, about 10 feet (3 meters) up. When I approached to get a better look, I saw that the bees were walking toward and disappearing into a knothole. They were moving in! This black walnut tree had long been special to me—for its immense trunk, deeply furrowed and dark-brown bark, sprawling limbs, yellow-green leaves shaped like fern fronds, and nuts inside aromatic hulls that stained my hands dark brown—but now it was super special. It was a bee tree. Ever since I was a little boy, and had studied the drawing of a bee tree in A. A. Milne's book *Winnie-the-Pooh*, I had hoped to find a real bee tree. At last, today, I had.

I figured this bee tree would be fun to watch, and indeed it was. I went to it often that summer, to see what I could learn about the bees by watching them at the entrance to their home. I saw bees standing around the knothole. Are they guards? I saw bees, presumably foragers, flying out of the knothole. Where are they going? I wondered about the bees' nest hidden inside the thick limb. Where exactly are the beeswax combs? What do they look like? How much honey is in them? One time, I lugged my father's heavy wooden stepladder to the tree to watch the bees close-up. I didn't have a beekeeper's veil, so I didn't dare get close enough to peer straight into the knothole. I did, though, get close enough to watch bees flying home with loads of pollen attached (somehow) to their hind legs. Other bees stood nearly still in the knothole, facing outward with their front legs and antennae raised. They looked extremely alert, so I guessed they were standing guard.

I remember wondering whether the knothole (Fig. 1.1) provided a big enough "breathing hole" for the thousands of bees living in the tree cavity. How come they don't suffocate? A few days later, I saw something that gave me a clue: about a dozen worker bees stood side-by-side along the bottom of the knothole, with their heads pointing into it, their bodies hunched over, and their wings whirring so fast that they were nearly invisible. It was a beautiful sight. The bees' light-brown bodies stood out against the tree's dark-brown bark, and each bee's wings hummed steadily, as if they were battery powered. These little fanners were still going strong when I stopped watching them about 10 minutes later. I had no doubt that they were expelling air from their home, but I could only guess *why* they were doing this. To cool it? To ventilate it? Perhaps both?

My mother must have noticed that I enjoyed watching the bees living in the black walnut tree, because next Christmas my parents gave me a

FIG. 1.1. The knothole in the lowest limb of the big black walnut tree near my boyhood home, where I first enjoyed watching honey bees up close. This nest entrance is approximately 1.5 inches (4 cm) in diameter.

book titled *The Makers of Honey*, written by Mary Geisler Phillips, a professor at Cornell University. It is a lovely, 164-page book, and I greatly enjoyed the hours I spent poring over it. The writing was the right "speed" for me and the finely crafted, scratchboard drawings in each chapter, by Elizabeth Burckmyer, delighted me. (They still do.) In Chapter 7, titled "Odd Jobs for Young Workers," Professor Phillips explains that one of the jobs of young workers is to be a fanner, and that "these fanners are air-conditioners . . . who keep the hive ventilated and at the right temperature." This description of fanners as "air-conditioners" satisfied my curiosity at the time. I suppose it also primed me to explore the behavior of these bees more closely when I got older.

My closer look at fanner bees came ten years later, in the summer of 1973. This was the summer following my third year as an undergraduate student at Dartmouth College, in New Hampshire. By then, I had taken courses in biology, chemistry, physics, and math, so I had picked up a fair amount of book knowledge and laboratory skills that I figured would be useful for what *really* interested me: studying honey bees. Also, by 1973 I had worked for three summers at the Dyce Lab for Honey Bee Studies at Cornell University (Fig. 1.2). The lab's director in those days was Professor Roger A. Morse. Everybody called him "Doc." Besides mowing the lawn, painting hives, and sometimes assisting Doc's graduate students and visiting scientists with their projects, I helped with the beekeeping. This was my favorite part of the job because it involved going to various apiaries and working with the bees. I continued to do laboratory chores in 1973, but that summer Doc said that I could devote some of my paid work-time to conducting a study of my own, on nest ventilation by fanner bees. Doc was a pretty gruff guy, but he was also supportive of his students . . . so long as they worked hard.

What spurred me to look closely at nest ventilation was something that I had seen back in September 1972, a few days before I would return to Dartmouth for the fall semester: rows of worker bees were fanning steadily at the entrances of my two hives *on a chilly evening of a rainy day*. This puzzled me because I figured that, given the conditions, these colonies didn't need fanners to cool the nest or to "ripen" fresh nectar into honey. (No colony has fresh nectar at the end of a rainy day.) I also figured that these fanners might, however, be ventilating their crowded home to avoid asphyxia. If so, then worker bees must be sensitive to either a lack of oxygen or an excess of carbon dioxide inside their nest.

Two days later, I did a simple experiment at Dyce Lab to see if a lack of oxygen stimulates worker bees to become fanners. The setup was easy. First, I moved a hive that housed a strong colony to a spot outside one of the laboratory's windows. Next, I drilled a quarter-inch diameter hole in the rear of this hive's upper box ("hive body") and inserted a glass tube through this hole so that that it poked into the center of the hive. Then, using a long rubber hose that snaked out the window, I connected the glass tube to a tank of compressed nitrogen inside the lab. The ex-

FIG. 1.2. The Dyce Laboratory for Honey Bee Studies at Cornell University, in summer 1998. Artist: Margaret C. Nelson.

periment started around 10:00 p.m. The air was cool, all the bees were at home, and I saw no fanners at the hive's entrance. I opened the tank's valve to send a gentle stream of nitrogen into the hive to displace the normal, oxygen-rich air inside. I figured that if worker bees are stimulated to start fanning by sensing a lack of oxygen, then what I was doing should elicit a strong fanning response. But it didn't. I neither saw nor heard any fanning bees. Displacing the oxygen from this colony's hive did, however, eventually narcotize the bees. I revived them by shutting off the gas and opening the hive. This experiment left me keen to find out how honey bees would respond if I sent a gentle stream of carbon dioxide into their home, but this follow-up experiment had to wait until the following summer.

What I could do in the meantime was find out what, if anything, previous bee researchers had reported on this subject. Back in the early 1970s, it was impossible to make a thorough search of the scientific literature—the launch of Google Scholar was 30 years in the future—but a kind librarian at Dartmouth helped me find two articles that contained information on the responsiveness of honey bees to carbon dioxide. The first was published in 1941 by a Dutchman, E. H. Hazelhoff. His report appeared in a Dutch beekeeping magazine, the *Maandschrift voor*

Bijenteelt [Monthly Journal for Beekeeping]. I could not locate a copy of this magazine and I could not read Dutch. So, all I knew about E. H. Hazelhoff's study was what a British entomologist, Dr. C. Ronald Ribbands, had written about it on page 212 in his 1953 book *The Behaviour and Social Life of Honeybees*: "Hazelhoff (1941) found that fanning commenced within one minute of the introduction of a stream of carbon dioxide in the hive." This finding intrigued me, for it was very different from what I had seen when I had introduced a stream of nitrogen. (Note: I remember thinking at the time [in November 1972] that E. H. Hazelhoff must not have done a rigorous study, for if he had then he would have published his report in a scientific journal, not a beekeeping magazine. At the end of this chapter, I will explain that I was *dead wrong* about the quality of Hazelhoff's study.)

The second article that I found on carbon dioxide in relation to honey bees was published in 1964 by a neurobiologist, Dr. Veit Lacher. He had worked in the Department of Comparative Neurophysiology in the Max Planck Institute in Munich, and he had published his study in a highly respected scientific journal, the *Zeitschrift für vergleichende Physiologie* [Journal of Comparative Physiology]. Lacher had made a detailed study of the sensitivity of the olfactory cells on the antennae of worker honey bees, and one of his discoveries was that some of these cells are sensitive *specifically* to gaseous carbon dioxide (CO_2). He reported that the threshold concentration for the response of these CO_2-sensitive antennal cells is about 0.50% (5,000 ppm) (Fig. 1.3). This is far above the level found in the atmosphere. Back in the 1960s, the atmospheric concentration of CO_2 was about 0.03% (300 ppm); today it is about 0.04% (400 ppm). I figured, though, that the 0.50% response threshold of the bees' CO_2-sensitive antennal cells might be just right for monitoring the gaseous CO_2 level inside the crowded nest of a honey bee colony.

In addition, Lacher reported a curious feature of the CO_2-sensitive cells: they keep firing for as long as they are stimulated. In the lingo of neurobiologists, these sensory cells have a "tonic response." This told me these cells could function very nicely as detectors of a dangerously high

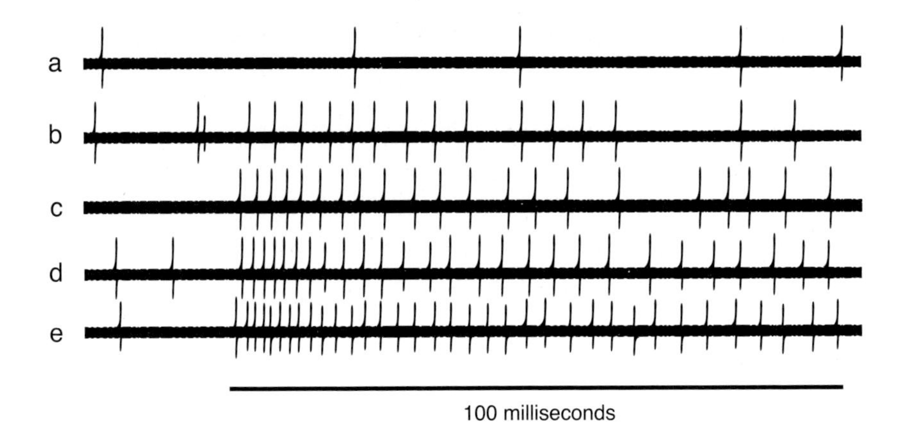

FIG. 1.3. Records of the electrical activity of an odor receptor cell (sensory neuron) in a worker bee's antenna that is sensitive specifically to CO_2. The top recording (a) shows the low rate of firing by this neuron when the antenna was exposed to air with 0.03% CO_2 (i.e., fresh air). The lower recordings (b–e) show how this odor-receptor cell started to fire more rapidly as soon as air with 0.5% CO_2 passed over it (line b) and increasingly so when exposed to air with 2%, 5%, and 10% CO_2 (lines c–e). In all cases, the cell fired steadily as long as air with elevated CO_2 passed over it. The bar at the bottom is a time scale.

level of CO_2 in the air inside a honey bee colony's home. As we all know regarding the smoke detectors in our homes, it is important that they keep sounding as long as there is danger.

Now I knew what I needed to do: (1) measure the gaseous CO_2 levels in bee hives, to understand what threats (if any) honey bees face from asphyxia by excessive CO_2; and (2) describe the response of worker bees to a high level of CO_2 inside their home. Would I see rises (and falls) in the gaseous CO_2 level in hives, and would I see corresponding rises (and falls) in the number of bees fanning their wings, as reported by Hazelhoff?

To address the first topic, I would need a sensitive CO_2 analyzer. I did not have one, but I figured that I could build one using an electronic sensor—a thermal conductivity (TC) cell—of the sort found in a gas chromatograph. I knew from my courses in organic chemistry that a TC cell works well for measuring tiny amounts of chemical compounds in samples of gases and (vaporized) liquids. I knew, too, that TC cells respond especially strongly to CO_2. Furthermore, I knew that a former graduate student in the Department of Entomology at Cornell, Dr. Larry J. Edwards, had built a sensitive CO_2 gas analyzer with a TC cell to study how insects breathe, and that he was now an assistant professor of entomology at the University of Massachusetts at Amherst. So, I mailed Professor Edwards a letter in which I explained my interest in his studies and requested a meeting to get his advice for my project. He sent a letter back, telling me when I could visit him at his laboratory in Amherst, Massachusetts. This was only about 100 miles (160 kilometers) south of where I was in Hanover, New Hampshire. I did not have a car, and getting to Amherst by bus was complicated, so I visited Professor Edwards by hitchhiking down (and back) on Interstate 91 in mid-December 1972. We talked, and when he understood my project and saw that I was pursuing it seriously, he did something that amazed me: he handed me the TC cell that he had used for his own studies (a GOW-MAC 133 Thermistor Cell). I will never forget his spontaneous act of generosity.

The next step was to build the electronic circuit that would convert the outputs of the TC cell into a voltage that could be fed into a strip chart recorder to produce a hard copy readout of the amount of CO_2 in an air sample. I wrote to the manufacturer of the TC cell, in Bethlehem, Pennsylvania, for information, and somebody there was kind enough to send me the electronic circuit's design (a "Wheatstone bridge"). I still have those instructions. Then a friendly professor in the Department of Physics at Dartmouth, Dr. William (Bill) Doyle, helped me build the circuit. So, when I returned to the Dyce Lab the following June, I had with me a homemade, but sensitive CO_2 analyzer. It enabled me to measure, with 1-microliter precision, how much CO_2 was in an air sample. Given this sensitivity, and given that I would work with small (25-milliliter, or about 1.5-cubic-inch) samples of air drawn from my study colonies, I knew that my measurements of the CO_2 levels in my samples would have a precision of 0.004%. As we shall see, this level of precision was sufficient.

This became clear as soon as I began measuring the gaseous CO_2 levels inside the hives of bees. I knew, of course, that honey bees live crowded together in their homes—tree cavities and bee hives—so I expected to find inside those homes levels of CO_2 that were higher than the 0.03–0.04% level of CO_2 in normal air. I was astonished, however, when I made my first measurements and learned how incredibly "stuffy" it can get inside a bee hive. To make these measurements, I set up two colonies side-by-side outside the Dyce Lab. Each colony was housed in a 10-frame Langstroth hive. One colony was large, so that in the evening (when all of its bees were at home) its adult bees covered thickly all 10 frames of comb in its hive. This colony's hive also contained many thousands of immature bees; eggs, larvae, and pupae nearly filled the cells in 7 of its 10 frames of comb. The other colony was smaller. Its adult bees covered only 5 of its 10 frames of comb when everybody was at home, and it had only 3 frames of comb whose cells held brood. To collect air samples from the centers of these hives, I installed in each one a glass tube that extended to the hive's center and poked out its rear wall. I used a rubber suction bulb to pull air samples (held in small glass flasks) from inside these hives.

I made my first measurements of the CO_2 levels in these two hives in June 1973. I did so by extracting, and then immediately analyzing, an air sample from each hive once an hour, from 9:00 p.m. on June 21 to 9:00 a.m. the next day. I worked at night because I wanted to see how stuffy it gets inside a hive when the entire colony is at home. The results, shown in Figure 1.4, surprised me in two ways. First, I found that the average CO_2 levels inside the hives of the large and small colonies were 0.55% and 0.92%, respectively. These readings were approximately 20 and 30 times higher than in the fresh air outside the hives. Yikes! This showed that the homes of honey bees can be extremely stuffy, at least by human standards. Breathing air with a CO_2 level of just 0.50% can make us quite drowsy. (This is why 0.50% CO_2 is the recommended limit for workplaces in the United States.) Second, I found that the CO_2 level varied far less in the hive of the larger colony (range: 0.39%—0.72%) than in the hive of the smaller colony (range 0.33%—1.77%).

After I took each air sample, I looked (using a flashlight) for fanners in the entrance of each hive. I am glad I did, because these inspections

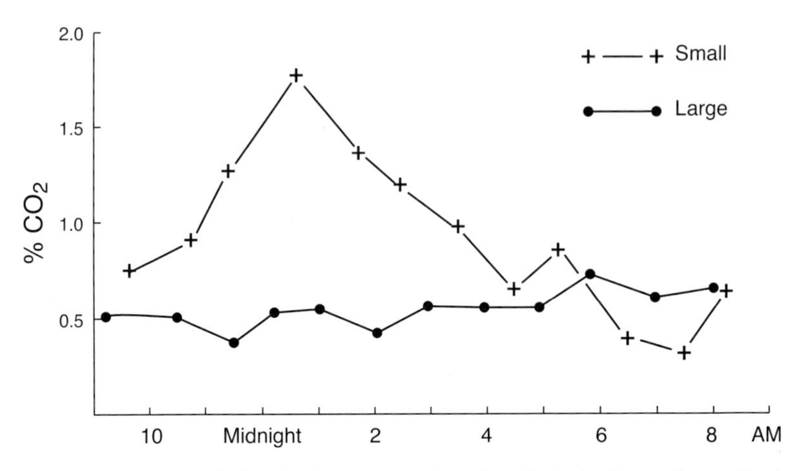

FIG. 1.4. Comparison of the levels of gaseous carbon dioxide in the hives of a small colony and a large colony throughout a cool night.

explained a lot about the difference in average CO_2 level between the two colonies. At the large (and crowded) colony, I *always* saw bees fanning at the entrance and heard the "roar" of more fanners working inside the hive. Two recent, high-tech studies of nest ventilation by honey bee colonies, by Jacob Peters and his colleagues at Harvard University, have shown that when workers stand at the nest entrance and ventilate their home, they use a special wing-fanning behavior (Fig. 1.5) that is different from the wing-flapping motions they use for flight. For example, the frequency and amplitude of the bees' wing movements are markedly different when they are fanning (174 Hz and 118°) than when they are flying (227 Hz and 87°). These studies also found that the stream of air that shoots from a hive's entrance can have a velocity of more than 10 feet per second (3 meters per second), which is 6.8 miles per hour (11 kilometers per hour). This is powerful ventilation!

At the smaller (and less crowded) colony, however, I *never* saw or heard fanners. I suspect that this colony was struggling to keep its brood warm, so was unable to keep its nest well ventilated throughout the cool night. This was certainly the situation between midnight and 2:00 a.m., for then its CO_2 level rose above 1.5%, which is high enough to make a human breathe fast and hard. I looked the next morning for dead bees in front

FIG. 1.5. Bees fanning their wings at their hive's entrance on a hot day.

of this hive, but I found only four—no more than usual. This showed me that honey bees can tolerate levels of CO_2 that are very stressful for human beings. This experiment also revealed something that, if you are a beekeeper, you may find surprising: when the outside air is cool, the air quality inside the hive of a small colony can be much poorer than that inside the hive of a large colony. This is because a small colony sometimes struggles to keep its home both properly warmed and suitably ventilated.

Besides the experiment just described (and various replicates of it), I did another experiment to check the accuracy of what the British biologist, C. Ronald Ribbands, had written when he summarized the work of E. H. Hazelhoff: "Hazelhoff (1941) found that fanning commenced within one minute of the introduction of a stream of carbon dioxide in a hive." For this study, I used a small observation hive that I built and then set up inside a heated room at Dyce Lab. I connected this hive to a tunnel through the building's wall, so the bees living in my hive could go outside and come back in as they wished. I stocked this hive with one frame of comb that contained brood, pollen, and honey, and that was covered with some 2,000 worker bees, one queen, and about a dozen drones. When I built this hive, I gave it two tubes, one at the top of the

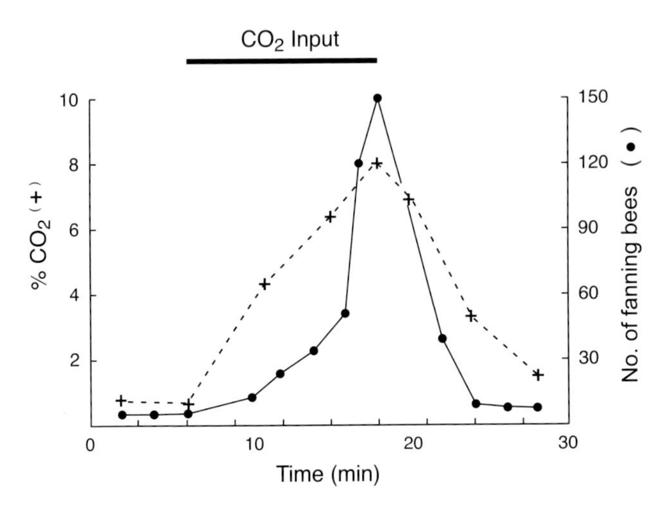

FIG. 1.6. Fanning response to induced rise in the carbon dioxide level in the air inside an observation hive. The numbers of fanning bees shown are based on counts of these bees inside the hive.

comb for introducing a gas (carbon dioxide or nitrogen), and the other halfway down the comb, for taking air samples from inside the hive. To perform a trial, I waited until it was night (when all the bees were at home); then I began extracting samples of the air in this hive and making counts of the bees in it that were fanning their wings. Once I had determined the baseline levels of CO_2 and of the bees' fanning, I introduced a light stream of either carbon dioxide (for about 10 minutes) or nitrogen (for about 30 minutes, long enough to displace the oxygen and narcotize the bees).

The results for the CO_2 trials—shown for one trial in Figure 1.6—confirmed Hazelhoff's report. Introducing a gentle stream of CO_2 stimulated strong fanning. But introducing a stream of nitrogen, even a strong one, did not stimulate fanning. This showed me that the bees' fanning response in the CO_2 trials was triggered by the CO_2 per se, not simply by the disturbance of introducing a gas.

I like very much the phrase "a canary in the coal mine," which refers to an early indicator of some danger or failure. Coal miners took caged canaries into mines as sentinels to tell them when the level of carbon monoxide or carbon dioxide had grown dangerously high (from the slow but steady oxidation of coal exposed to air). If the canaries stopped singing, then the miners knew that they needed to get out! In this chapter, we have seen that honey bees, like coal miners, face a danger of asphyxiation from carbon dioxide, and that they, too, have sentinels—the CO_2-sensitive olfactory cells on their antennae—to tell them when to take action to deal with the problem. We have seen, too, that for honey bees, as for coal miners, the best solution to this problem is to boost the ventilation of their workplace.

Postscript: In 2018, while writing my book *The Lives of Bees*, I discovered several important things about E. H. (Engel Hendrik) Hazelhoff. First, I learned that he was not a beekeeper, but rather a distinguished professor of zoology in the Department of Natural Sciences at Groningen University in the Netherlands. Second, I learned that he died unexpectedly and young (age 45 years) on 30 September 1945. Third, I learned that the study of bee hive ventilation that he reported in 1941, as a series of four articles in the Dutch beekeeping magazine *Maandschrift voor Bijenteelt*, was republished posthumously in English in 1954, as a 26-page paper in the scientific journal *Physiologica Comparata et Oecologia*. It shows that Hazelhoff made a meticulous study of what he called "the social regulation of the respiration of a honey bee colony." He built a sophisticated hive which he could cool or heat, monitor for internal temperature and gaseous carbon dioxide level, and ventilate at different levels by adjusting the rate of airflow through it. Using this hive, he conducted a rigorous study of the conditions inside a hive that stimulate ventilation fanning by bees: either a high temperature or a high concentration of gaseous CO_2. There is no doubt that Engel H. Hazelhoff was the first person to show that a high level of gaseous CO_2 in the home of a honey bee colony triggers fanning by its worker bees.

CHAPTER 2

Forest Homes

The history of beekeeping is mainly a story of people seeking better ways to get honey from colonies of honey bees. Half of this story is about *how beekeepers manage* the bees living in their hives, and the other half is about *what tools they use* for their work. Their most important tool, by far, is the hive. Beekeepers have devised many kinds of hives—clay pipes, hollowed logs, and the coiled-straw domes called skeps—but the movable-frame hive invented by Lorenzo L. Langstroth in 1851 is probably the most important hive worldwide. A Langstroth hive consists of a stack of rectangular boxes, each of which holds 8 or 10 carefully manufactured, accurately spaced, and interchangeable wooden frames, which hold the beeswax combs that the bees inhabiting the hive have built. Langstroth hives are readily moved, are easily enlarged (or reduced), and can be smoothly opened when beekeepers inspect their colonies.

Because nearly every colony of honey bees that one sees these days is living in a manufactured hive, it is easy to think that honey bees are domesticated animals. But this is not so. The honey bees living under human management do not differ in behavior from those living in a wild state, such as in a tree cavity, rock crevice, or abandoned hive. Why is this? Although honey bees have been the subjects of breeding programs for various traits—better disease resistance, reduced defensiveness, and improved pollination of certain crop plants, for example—few of these programs have been maintained, so most of the artificially selected lines have not persisted. Moreover, in most places in the United States (and many other countries), the queens and drones from beekeepers' colonies mate with their counterparts from wild colonies, and this gradually erases any effects of artificial selection. So, it is unlikely that the long association between honey bees and human beings has modified the honey bee's fundamental traits. This is why, for example, a swarm of honey bees that escapes a beekeeper's hive can move into a hollow tree and live there all on its own for many years.

The mystery that captivated me as a 10-year-old, of the wild colony's home inside a thick limb of "my" black walnut tree, stayed with me as I grew older, became a beekeeper, and read books on the biology of honey bees. I learned that archaeological findings, such as cave paintings in Spain, show us that honey hunters have plundered the nests of wild colonies for tens of thousands of years. I also learned that the nests of wild honey bee colonies had never been carefully described. Amazing! I was delighted to discover this gap in our knowledge of the basic natural history of our familiar honey bee, *Apis mellifera*, and I decided to try to fill it.

I made this decision in the summer of 1974, when I was 22 years old and was looking ahead to starting my studies at Harvard University to earn a Ph.D. in biology, with a focus on insect social behavior. Every Ph.D. student needs to choose an unsolved problem for his or her thesis research work, and I chose the mystery of nest-site selection by honey bees. I did so partly to pursue my long-standing curiosity about the natural homes of honey bees, and partly because I was keen to extend a fascinating investigation reported in 1955 by a German biologist, Martin Lindauer. He described how the scout bees in a honey bee swarm find potential sites for their colony's new home and then "advertise" them by performing waggle dances on the surface of the swarm. A swarm is a group of approximately 12,000 worker bees and their queen that have left their home to establish a new colony. Usually, these bees assemble in a beard-like cluster that hangs for a day or so from a tree branch, or other such site, near their old home. It is from here that the nest-site scouts usually conduct most of their search.

At first, the waggle dances performed by the scouts indicate several potential home sites, but after a few hours or a few days, these dances indicate just one site, generally the best of the various options that they have discovered. Lindauer's investigation showed that the nest-site scouts are evaluating potential home sites and then advertising them accordingly, but his work did not reveal what makes a site attractive to these small house hunters. What exactly do they seek in a dwelling place? My goals were clear: (1) describe the nesting sites and nests of wild honey bee colonies, (2) determine what makes a cavity in a tree or building desirable to nest-site scouts, and (3) investigate how a nest-site scout inspects a potential home site.

I made progress toward the first two goals by working at the Dyce Laboratory for Honey Bee Studies at Cornell for much of each summer in 1975, 1976, and 1977. The laboratory's director, Doc Morse, welcomed me back and let me use the things in his lab that I needed for my studies: various hand tools (chain saw, steel wedges, sledgehammer, and peavey, i.e., a lumberjack's heavy-duty lever with a steel tip), the woodworking shop, and a pickup truck. Doc also arranged the help of Herb Nelson, a member of the technical staff of the Department of Entomology (Fig. 2.1). Herb had worked as a logger in the State of Maine. He showed me how to safely fell the trees that housed wild colonies, and then get the heavy logs enclosing the colonies' nests out of the woods and onto a truck. Without Herb's skilled help, Doc and I could not have done this study.

I knew of three bee trees on Cornell land, and to find more, I put a "want ad" for trees housing live colonies of honey bees in the local paper, the *Ithaca Journal*. It read, "**BEE TREES** wanted. Will pay $15 or 15 lb. of honey for a tree housing a live colony of honey bees." Most of the folks who responded to my ad wanted their payment in jars of honey, which was helpful. My ten colonies of bees at my parents' house in Ellis

FIG. 2.1. My friend, Herb Nelson. A classic "Mainer."

Hollow produced a big honey crop in 1975, so that summer I had lots more honey than money.

In all, I secured permissions from 36 landowners to examine, and in some cases to cut down, the bee trees on their properties. This ad brought me in touch with folks living in some lovely, out-of-the-way places up in the hills around Ithaca. The first person was Prudy Mix, the wife of a dairy farmer whose farm was only a few miles from my parents' house. She was keen to get some honey, so she took me up a narrow dirt road through the woods to her family's hunting camp and showed me where honey bees were living in a sugar maple tree (*Acer saccharum*) (Fig. 2.2). The second person was Gussie Gaskill, a scholar of Chinese language and literature at Cornell University. She had a little summer house atop a cliff overlooking Cayuga Lake and had noticed honey bees living in a butternut tree (*Juglans cinerea*) nearby. The third was a man who was living by himself in a log cabin beside a pond in Danby. While he was showing me the honey bee colony that he had found, living in an elm tree (*Ulmus americana*), I noticed he looked sad, so I asked "Excuse

FIG. 2.2. The first bee tree whose nest I dissected. It was a 120-year-old sugar maple that was nearly dead. The knothole in the left fork was the nest entrance.

me, but are you OK?" He responded that he had learned a few days before that his son had been killed in Vietnam. We walked to his cabin and sat on the porch. He talked about his son. I listened, tried to say comforting things, and thought about my oldest brother, Dave, a Marine Corps pilot who was flying a "Huey" helicopter in Vietnam. I learned about more things than just bee trees from this project.

The first steps in studying the home of each wild colony were to photograph the bee tree and then measure and record various details about the nest's entrance: its size, shape, height, and compass direction. Herb and I then felled the tree, cut out the section that housed the bees' nest, rolled and wrestled it into the pickup truck, brought it back to Dyce Lab, and finally got it into the workshop. Once we had one of these bee-tree "logs" safely indoors, I split it open to expose the bees' nest. Then I carefully took it apart, comb by comb (Fig. 2.3). It was fascinating to dissect the nests of these wild colonies and thereby discover their secrets. It was also sobering, for I had killed each colony—very quickly, using cyanide gas—before sunrise on the day we felled its tree. I did this so Herb and I could collect these nests without getting heavily stung, and so I could count how many workers and drones were in each colony. Even after 47 years, these awkward memories still elbow themselves to the front of my mind when I think about bee trees. No beekeeper likes to harm, let alone kill, bees.

This investigation of the natural nests of honey bees was not rocket science, but it was revealing science. It disclosed many striking differences between the forest homes of wild colonies and the apiary homes of managed colonies. Besides the obvious difference of home *spacing*—far apart in the woods vs. close together in apiaries—there were many differences in home *design*: size and height of nest entrance, size and shape of nest cavity, thickness of cavity's walls, and many more (see Table 2.1). These glaring differences began to open my eyes to the many ways that conventional beekeeping practices alter the lives of honey bees in ways that benefit us but can harm the bees: more risk from disease (from crowding colonies into apiaries), more energetically costly nest thermoregulation

FIG. 2.3. The nest inside the bee tree shown in Fig. 2.2. The nest entrance is on the left side, about two thirds of the way up the cavity. The light-brown combs at the top contain honey and the dark-brown combs below contain brood.

(from using thin-walled hives that provide poor insulation), and less reproductive success (from inhibiting swarming and drone rearing), to mention just a few.

Looking back, I see that this study of the natural homes of honey bees was not just an early stage in my journey into science, but also in my thinking about how to be a bee-friendly beekeeper. For example, I now prefer to use hives whose walls are well insulated with foam board. This

TABLE 2.1. Comparison of the homes of wild and managed colonies

Property	*Wild colony*	*Managed colony*
Spacing	far apart, in forest	near others, in apiary
Structure	hollow tree	wooden box
Wall thickness	average, 6 inch (15 cm)	average, 0.75 inch (2 cm)
Propolis on walls & floor	yes	no
Cavity shape	tall & cylindrical	squat and boxy
Cavity volume	average, 1.4 cu. ft (40 liters)	often, 3.0–6.0 cubic feet (80–160 liters)
Entrance height	average, ca. 21 feet (6.5 m)	average, < 1 foot (0.3 m)
Entrance area	average, 1.5–3.0 sq. inches (10–20 sq. cm.)	average, ca. 12 sq. inches (ca. 75 sq. cm.)
Entrance location	near bottom of cavity	at bottom of hive
Entrance direction	usually south-facing	often south-facing
Number of combs	8 maximum	10 maximum
Comb area	ca. 3700 sq. in. (2.4 m^2)	2400 sq. in. (1.5 m^2) per box
Drone comb	17% of comb area	usually, very little
Wall-to-wall combs?	Yes	No
Deep-celled honey combs?	Yes	No

raises their R-value from 1 (that of a standard hive made with 3/4-inch (1.9-cm) pine boards) to about 8 (that of a tree cavity whose walls are 6 inches (15 cm) thick). This means that, summer or winter, heat passes through the walls of my hives 8 times more slowly than before.

The work of describing the natural nests of honey bees showed me what their homes are like when they live out in the forests around Ithaca, but it did not tell me which features of their homesites are important to them and which ones are incidental to living in hollow trees. For example, it might be that their nest entrances tend to be smallish knotholes because the nest-site scouts for swarms instinctively choose nesting sites with small entrances so their homes are snug and easily defended. On the other hand, a tall and cylindrical nest cavity may be the norm simply

FIG. 2.4. Painting of a log hive occupied by bees, with Maasai watching the bees. Purchased by Roger A. Morse in 1972 in a market in Kenya. It hung in his office at the Dyce Lab. The artist is unknown.

because this is the most common shape of cavities in tree trunks. Doc and I needed to ask the bees about their nest-site preferences.

Our plan for doing so came from something that we knew about beekeeping in East Africa and South Africa. Beekeepers in these places acquire swarms by hanging bait hives—hollowed-out logs—horizontally in trees and waiting for swarms to occupy them (Fig. 2.4). Birdwatchers the world over do something similar: putting out bird houses to attract cavity-nesting birds.

I had not heard or read of beekeepers in North America putting out bait hives to catch swarms. But I supposed that this might work, and that if it did, then Doc and I could ask the bees about their housing preferences by putting up groups of two or three bait hives (nest boxes), with the boxes in each group identical except for one property, such as entrance size. If swarms found our boxes, and if they consistently occupied those with certain attributes, then they would reveal to us their housing preferences.

I did a small-scale pilot study in 1974 in which I nailed six boxes onto trees along the roads in Ellis Hollow, near where I grew up. Each box offered 1 cubic foot (28 liters) of nesting space, had an entrance hole that was 1.75 inches (4.5 centimeters) in diameter, and was mounted about 15 feet (5 meters) off the ground. These boxes looked like bird houses on steroids, except that each one had some chicken wire stapled over its entrance hole to keep birds out and let bees in. By the end of July, swarms had moved into three of my six nest boxes, so Doc and I knew that our plan for testing the nest-site preferences of swarms was likely to work. There were swarms to catch and they would move into artificial nesting sites.

Indeed, our plan worked surprisingly well. Each summer for the next three years (1975–1977), Doc and I mounted some 250 nest boxes along roads in the countryside around Ithaca. We deployed them in groups of either two or three boxes spaced about 30 feet (10 meters) apart (Fig. 2.5), and in each summer over half of these nest-box groups had a swarm occupy one of its boxes. In each group of nest boxes, there was one box whose properties (entrance area, cavity volume, etc.) matched those of a *typical* nest site in nature, and another box (or two) that was (or were) identical to the first except in one property, the value of which was *atypical* for a natural nest site. This experimental design enabled us to test for a preference in the property that differed between the boxes in a group. It required a mammoth effort in nest-box construction and

FIG. 2.5. Two nest boxes mounted on power line poles about 30 feet (10 meters) apart. These two boxes offered identical nest sites (same cavity volume and shape, same entrance direction and height, etc.) except that the one on the left had a larger entrance hole than the one on the right.

deployment. I spent most of December 1974 sawing, hammering, and painting in the woodworking shop at Dyce Lab to build the 252 nest boxes that we needed for our experiment. Then in April 1975, I spent a couple of weeks driving around in a pickup truck loaded with nest boxes and a ladder, looking for suitable pairs—or trios—of trees or power-line poles on which to mount them, and then doing so. Some Saturday mornings Doc joined me, and on these days the work started with a stop at the nearby Dunkin' Donuts. I learned many things from Doc, mostly about beekeeping, but also about stuff like the importance of taking time to enjoy simple pleasures, such as a cup of black coffee with a Maple Frosted Donut.

Our hard work was well rewarded. We captured 124 swarms in our nest boxes over the springs and summers of 1975, 1976, and 1977. This "trapping" success provided lots of colonies for my fellow graduate students working at Dyce Lab, but even more valuable were the discoveries it yielded. We learned that nest-site scouts have preferences regarding six properties of their new homesites (A > B means option A was chosen significantly more often than option B):

1. entrance area (2 square inches > 12.5 square inches),
2. entrance height above ground (15 feet > 3 feet),
3. position of entrance (bottom > top of cavity),
4. entrance direction (south > north),
5. cavity volume (1.4 cubic feet > 0.35 cubic feet),
6. preexisting combs from another colony (with > without).

I will not be surprised if further studies reveal that honey bees have still more real-estate preferences. For example, they may favor a thick-walled cavity (good thermal insulation and high thermal mass) over a thin-walled cavity (poor thermal insulation and low thermal mass). I should mention that we found no evidence that nest-site scouts prefer a tall cavity over a cube-shaped one with the same volume. This supported our hunch about why most of the wild-colony nests that we studied were tall and slender: simply because most of the cavities in hollow trees have this shape.

What is the adaptive significance of each of the nest-site preferences that we found? I can offer some speculations mixed with a few facts. A *small nest entrance* is probably beneficial because it is easily defended and because it helps make the nest cavity snug. A *high nest entrance* is probably good because it is less apt to be spotted by predators—such as black bears—than one at ground level. A *nest entrance near the cavity bottom* must help minimize heat loss from the nest by convection (warm air rising). And an *entrance facing south* provides a brightly lit, and solar-heated, takeoff and landing spot.

Why did the swarms spurn our small (0.35 cubic feet, 10 liters) and our large (3.5 cubic feet, 100 liters) nest boxes, but occupy our intermediate-size (1.4 cubic feet, 40 liters) ones? I think they avoided our small boxes because the nest-site scouts knew instinctively that these boxes were too small to be good home sites. Where I live, a colony needs 40 or more pounds (18+ kilograms) of honey to survive the six months of cold and flowerless days that we endure, from mid-October to about mid-April. If a swarm were to make its home in a 0.35-cubic-foot (10-liter) nest cavity, then it would have room for only about 20 pounds (9 kilograms) of honey stores, so probably it would starve over winter.

As for our large (3.5 cubic feet, 100 liters) nest boxes, I suspect that the nest-site scouts that inspected them judged them to be just too roomy. If a colony had occupied one of these big boxes, then probably it would have had difficulty in reproducing (swarming), because its bees might never feel crowded in such a spacious home. Beekeepers know that housing their colonies in large hives (by "supering 'em up") is essential for getting large honey crops. This is partly because large hives provide colonies with enough space to hold large stores of honey, and partly because large hives help prevent colonies from casting swarms.

The adaptive significance of a swarm preferring a nest site equipped with combs—built by a prior colony—seems clear. A colony enjoys a quick start and tremendous energy savings in setting up its new home if it occupies a site that contains a full set of combs. A few calculations make the energy-savings advantage clear. A typical bee-tree nest contains some 100,000 cells arranged in combs whose total surface area is about 3 square yards (2.5 square meters). Building this much comb requires about 2.5 pounds (1.1 kilograms) of beeswax. We know that the weight-to-weight efficiency of beeswax synthesis from sugar is about 0.20, so we can estimate that building the combs in a typical nest requires about 12.5 pounds (5.7 kilograms) of sugar, hence about 15 pounds (6.8 kilograms) of honey (which is about 82% sugar). This quantity of honey is about one-third of what a colony will consume over winter. Boosting a colony's honey stores for winter by 15 pounds (6.8 kilograms) by not spending this much honey on comb building must surely improve a colony's odds of surviving its first winter in a new home.

Besides learning about what is important to nest-site scouts, Doc and I learned about what is *not* important to them: entrance shape, cavity shape, cavity draftiness, and cavity dryness. Why don't scout bees care about these properties? I am not sure about entrance shape and cavity shape, but I am sure that honey bee colonies benefit greatly from having homes that are tight and dry. However, the colonies that occupied nest boxes we had made drafty or damp—to test the importance of cavity tightness and dryness—showed us that honey bees *can and will remedy* these two shortcomings. In contrast, a colony cannot modify the volume of its nest cavity, the height of the cavity's entrance, or the direction in which the entrance faces. So it makes sense that nest-site scouts take note of these factors when they inspect potential home sites, but show no signs of requiring their future homesite to be perfectly tight and dry.

I had rendered some of our nest boxes noticeably drafty by riddling their fronts and sides with quarter-inch (6-mm)-diameter holes spaced 3 inches (7.6 centimeters) apart, and I had made some other nest boxes obviously wet inside by dumping into them two quarts (about two liters) of waterlogged sawdust just before I nailed them onto their trees. Moreover, I kept the sawdust in these "wet boxes" soggy by squirting in another quart of water whenever I checked one and found it still unoccupied. When I took down the occupied nest boxes, I discovered that every swarm that had moved into a drafty nest box had made its new home draft-free. Honey bees can caulk with propolis—a mixture of tree resins and beeswax—any cracks that could let in wind and rain, and the bees had plugged the holes I'd made in the nest box walls with blobs of this "bee glue." Likewise, every swarm that had moved into one of the wet nest boxes had rendered its new home perfectly dry by removing all the soggy sawdust. Lesson learned: honey bee swarms will accept fixer-uppers, and they are willing to do repair work.

CHAPTER 3

Homesite Inspectors

While working as a summer helper at the Dyce Laboratory in the early 1970s, I discovered in its library a small, plainly written, and splendid book titled *Communication among Social Bees*. Its author is the German biologist Martin Lindauer, and it had been published in 1961. Lindauer was the most distinguished student of Karl von Frisch, the Nobel laureate at the University of Munich who deciphered the amazing waggle dance of honey bees. Lindauer had a special knack for noticing curious little things about honey bees that led him to big discoveries, especially about their social behavior. The book that I found was based on a special set of lectures that Lindauer had delivered—the Prather Lectures for 1959—at Harvard University. Each lecture described an investigation with honey bees that he had made in the late 1940s or the 1950s. These are the years when Lindauer was a Ph.D. student, and then a colleague, of Karl von Frisch.

Martin Lindauer was born in 1918 and grew up in the small village of Wäldle, which is in the foothills of the Bavarian Alps, south of Munich. He was drafted into the German army upon graduating from high school in 1939. In July 1941, he was badly injured during a battle in eastern Ukraine in which German troops attacked a fortified defensive line of Soviet forces (the "Stalin Line") that were trying to halt the advance of the German army toward the city of Stalingrad (now called Volgograd).

Lindauer was transported to a hospital in Munich, where he gradually recovered. In January 1942, he received permission to leave the hospital and explore the city. What he wanted to see most was the university, for he had been a star student in his high school. The doctor who tended him recommended that he attend a lecture by Karl von Frisch, and he did so. The lecture he watched was part of von Frisch's winter semester course, General Zoology. Von Frisch talked that day about cell division, and he described his subject with beautiful charts and even a film. Lindauer came to this lecture from a war-torn world of material chaos and spiritual crisis, but as he listened to it he caught a glimpse of a new world, one devoted to revealing the secrets of life. This was a world that offered him hope of again finding meaning in his life.

In the early 2000s, I enjoyed several long stays at the University of Würzburg, in Germany, where my onetime Harvard *Doktorvater* (Ph.D. advisor), Bert Hölldobler, had succeeded Martin Lindauer as the Chair of the Institute of Behavioral Physiology and Sociobiology. This gave me the opportunity to meet Martin Lindauer, and we became good friends. Eventually, I collaborated with a student at this university, Susanne Kühnholz, to interview Lindauer at length about his life and work, in preparation for writing his biography. During one of our interview sessions, he shared with Susanne and me the memory that when he watched Karl von Frisch standing there talking about cell division, two things were imprinted on him: first, a new world of humanity, one in which humans create rather than destroy; and second, an encounter with science, an endeavor where humans use truth rather than lies. Lindauer also shared with us that during this lecture, he resolved to study biology at the University of Munich. Soon he did so, starting in the summer of 1943, following his discharge from the German army as a severely wounded soldier.

Lindauer's small book about honey bees has five chapters, one for each of his lectures at Harvard. The chapter that intrigued me most was Chapter 2, titled "Communication by Dancing in Swarm Bees." In it, he summarizes his discoveries about how the nest-site scouts in a swarm work together to choose their group's new home. Working out of what was left of the Munich Zoological Institute shortly after the end of World War II, he had patiently watched honey bee swarms and observed how the nest-site scouts, soiled with brick dust or chimney soot from examining cavities in the ruins of buildings, produced waggle dances to choose their new homesites. Lindauer's findings fascinated me, and I knew that I must read the scientific paper in which he reported his studies. So I went to the Mann Library at Cornell and made a photocopy of Lindauer's full account of his work, which had been published in 1955 as a 62-page journal article titled "Schwarmbienen auf Wohnungssuche" (swarm bees out house hunting). That his report was written in German stymied me for a while, but after taking a course in German, I was able to decipher it. It became clear that Lindauer's study had produced dozens of beautiful discoveries, but that, like all pioneering investigations, it had raised more questions than it answered. One was the question that I would focus on for my Ph.D. thesis: How do nest-site scouts assess a potential homesite? More specifically, what properties must a site have to be a first-class dwelling place, and how do the scout bees determine whether or not a site has these properties? In short, I wanted to find out how worker bees function as homesite inspectors.

My work on this matter began in August 1974. This was a few weeks before I would leave home to begin my Ph.D. studies at Harvard, and I had decided to attempt a small-scale, observational study at my parents' house. I brought home a swarm of some 10,000 bees and set it up on a wooden cross (i.e., an artificial tree branch) behind the house (Fig. 3.1). I induced the bees to cluster there by lashing to the cross the small cage in which I had confined the swarm's queen. The day before, I had nailed a bait hive about four feet (1.2 meters) up on a white pine that had sprung up in an abandoned field. This tree stood about 500 feet (some 150 meters) from where I set up the swarm. My bait hive was a simple structure: an empty, cube-shaped wooden box with a small entrance hole on its front side. I knew that I would be able to watch the scout bees doing

FIG. 3.1. The author, watching nest-site scouts on the surface of a swarm as they performed waggle dances to choose this swarm's new home. August 1974.

things on the surface of the swarm, but I wondered if I would be able to watch the scouts inspecting my bait hive. Would they find it and get interested in it?

I was lucky. My setup worked. Watching at the swarm cluster, I soon spied bees performing lively waggle dances and I knew, thanks to the codebreaking work of Karl von Frisch in 1944, that they were advertising a site that was nearby *and* in the direction of my bait hive. (By then, I had learned how to read the dances of honey bees. How one does this will be explained in Chapter 16.) So, I dashed to my bait hive and there I found several bees flying slowly around it, often hovering just an inch or two (3–5 centimeters) from it while facing it. Nest-site scouts! Note: the manner in which scouts examine the exterior of a potential homesite is distinctive. So, if you put up a bait hive, and you see bees behaving outside your hive as I have just described, then you can be sure that your bait hive is being examined by scout bees.

When I first watched this scouting behavior, I could see that these bees were inspecting the hive's exterior. But what items were on their "inspection list"? Drafty holes? Troublesome ants? Rotted wood that revealed structural weakness? I also saw bees land at the entrance hole and run inside; meanwhile, other bees scurried out, crawled around the entrance hole, and then popped back in. Then I heard what sounded like bees flying *inside* the box and bumping into its walls. Hmm. What is this all about? I didn't put paint marks on the bees visiting my box—because I did not want to disturb them—so I couldn't tell how long the interior inspections by these bees lasted, but I could tell from the way they dashed into and out of the entrance hole that they were very excited. Me too!

When I returned to the swarm cluster and studied the waggle dances being performed on its surface, I found that most were advertisements of my bait hive. Great! It looked like the scouts were on track to choose it for the swarm's home.

The next morning, I watched the swarm take flight and start moving to its new home. I then ran pell-mell to my bait hive, to watch the bees arrive there and move in. Never will I forget the sight that I saw that morning: a swirling ensemble of some ten thousand bees coming across the field, with sunlit wings sparkling against the dark background of the hedgerow trees they had just flown over. This cloud of bees soon reached the pine tree and began to condense on the front of my bait hive. I watched as the bees crawled inside with great "fanfare": dozens of bees were standing around the entrance hole, facing into it, and madly fanning their wings to disperse the lemon-scented come-hither pheromone from their Nasonov glands at the tips of their raised abdomens. If you collect a swarm by shaking it from a tree branch or bush onto the ground in front of an empty hive, then you, too, will see this lovely sight.

Observing this swarm made me eager to deepen our understanding of how the nest-site scouts in a swarm collaborate to select their future dwelling place, but I sensed that this line of study was not feasible. It would require getting all the scout bees in a swarm labeled for individual identification and then recording each one's behavior throughout the process of choosing the new homesite. I would do these things 20 years later, with sophisticated video recording equipment, but in the mid-1970s I was a novice researcher, so I needed to tackle a smaller and simpler problem. Therefore, for my Ph.D. thesis, I examined two parts of the nest-site selection process that I knew were interesting and I figured were tractable: What features do scout bees seek in a nest cavity? And how does a nest-site scout sense the size of a potential nest cavity?

Both lines of investigation required a blend of descriptive work and experimental work. In Chapter 2, we saw that solving the mystery of what honey bees seek in a nest site was a two-stage process: first, describe the natural homes of honey bees, and second, conduct choice experiments with nest boxes to find out which features of their homes reflect their "real estate" preferences. In this chapter, we will see that solving the mystery of how a nest-site scout senses the volume of a potential home site was also a two-stage process: first, describe the nest-site inspection behavior of scout bees, and second, conduct experiments to find out how this behavior helps a scout bee know whether a cavity is (or is not) sufficiently spacious.

Why did I focus on the mystery of how a nest-site scout senses the size of a potential nest cavity rather than, say, the height of its entrance? It was because I knew that having a sufficiently roomy nest cavity is a life-or-death matter for a colony of honey bees. Where I live, if a colony were to move into a cavity too small to hold at least 40 pounds (about 18 kilograms) of honey, then it would starve during a long, cold winter. Also, I had learned from dissecting the homes of wild colonies that they have a lower limit of nest-cavity volume: about 0.70 cubic feet (approximately 20 liters) (Fig. 3.2). This is just enough space for a set of beeswax combs that will hold 40 pounds (18 kilograms) of honey. Incidentally, this explains why in my tests of the housing preferences of honey bees, discussed in Chapter 2, no swarms moved into the smallest (10-liter, ca. 0.35-cubic-feet) nest boxes. So, several lines of evidence told me that

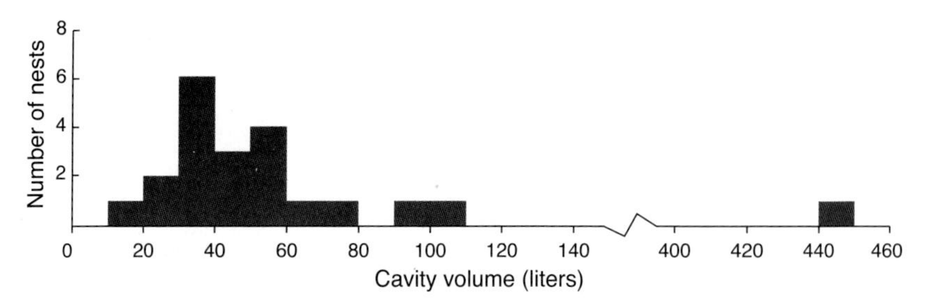

FIG. 3.2. Distribution of the nest-cavity volumes for 21 nests of wild colonies of honey bees living in tree cavities. It shows that in nature, the nest cavities of colonies are usually around the size of one deep Langstroth hive body (volume = 42 liters).

nest-site scouts pay close attention to cavity volume, and I wondered how they do so.

To solve this mystery, I had to leave the heavily wooded countryside around Ithaca and work someplace where there were few, if any, natural homesites for honey bees. One place that was highly promising was Appledore Island: a rocky, windswept, 95-acre (39-hectare) island that sits 7 miles (11 kilometers) out in the Atlantic Ocean off the coast of southern Maine. In 1966, Cornell University and the University of New Hampshire had established the Shoals Marine Laboratory on this island, and I knew of the Cornell professor who was its founding director, Dr. John M. Kingsbury. I went to his office over Christmas break in 1974, explained my Ph.D. thesis project, and asked him if I could spend several weeks next summer on Appledore to perform behavioral experiments with honey bees. He was intrigued by my project, and said I would be welcome. He understood that the vegetation cover on Appledore Island—tangles of blackberry brambles, wind-battered cherry bushes, thickets of poison ivy, and some grassy areas around the buildings (Fig. 3.3)—would force the scout bees in swarms to focus their attention on the nest boxes that I would provide. This was usually, but not always, what happened, as I will explain shortly.

I began my studies on Appledore Island in June 1975. My first goal was to describe the behavior of nest-site scouts when they inspect a potential nest cavity. To do so, I built a hut with a square opening (12″ × 12″,

FIG. 3.3. The treeless landscape of Appledore Island, Maine.

or 30 × 30 cm) in one wall and a lightproof door in another (Fig. 3.4). On the side of the hut with the square opening, I mounted a cube-shaped nest box that was open on one side. The open side of the nest box was fitted to the opening in the hut's wall, so I could peer into the nest box when I sat inside the hut. I covered the "window" into the nest box with a sheet of red Plexiglas. This enabled me to watch the scout bees inside the nest box but prevented them from seeing me inside the hut; the eyes of bees are insensitive to red light. The inner surfaces of the nest box were marked with a grid-coordinate system, so I could record how a scout bee moved around in the box while she was inspecting it. I set up my observation hut/nest box near the southwest corner of Appledore Island, and then I put a small swarm (fewer than 2,000 bees) on a wooden cross near the concrete tower from World War II that stands atop a high point on this island. Next, I labeled most of the bees in this swarm for individual identification using a method that was simple: using fine paintbrushes and bottles of Testors Pla enamel in a dozen colors, I put dots of

FIG. 3.4. *Top*: Interior of the nest box used to observe scout bees as they inspected a potential home site. *Bottom:* The nest box mounted against the opening on the side of my lightproof hut. A sheet of red Plexiglas normally covered this opening, but was removed when I took the top photo. Because bees do not see red light, I could watch them without disturbing them. I recorded the movements of individual scouts inside the box on audiotape by reference to the numbered squares.

paint—in various colors, in various places, and in various shapes—on the bees in my swarm. Then I sat in my hut and waited for a nest-site scout to arrive and show me how she would inspect the living quarters that I was offering. I crossed my fingers that my nest box was the best homesite for honey bees on Appledore Island.

After waiting at my hut for about an hour without seeing a scout bee approach my nest box, I plugged its entrance and dashed to the swarm to check on it. I was dismayed when I got there, for I saw several bees dancing excitedly, which told me they had found an attractive dwelling place that was not my nest box. Rats! Their dances also told me the direction and distance to their find. When I plotted its location on my map of the island, I saw that it was at, or around, the house of Rodney Sullivan, the lobster fisherman who lived on Appledore. This was bad news for me. It was also bad news for Rodney, who I had been told didn't "cotton to anybody poking around his place." I figured this would be especially true if the "anybody" was a party of house-hunting honey bees. So I sought the advice of Professor Kingsbury, who (thank goodness!) was on the island. He was both understanding of my situation and concerned about keeping good relations with his island neighbor. Soon, Dr. Kingsbury was introducing beeman Seeley to fisherman Sullivan. Rodney had already noticed my bees flying in and out of his chimney and had wondered how in the heck they got out to Appledore. After talking a bit about his chimney and my project, he and I settled on a plan: he would kindle a fire in his woodstove to evict my scout bees, and I would climb his steep and slippery (with gull poop) roof and tape some scrap window screen over his chimney's flue to prevent my bees from returning. Soon, the bees were out, the screen was on, and I was safely down. Success!

Eventually, a scout bee appeared at my observation nest box, and I recorded how she behaved when she was inside it. I did the same for several more scout bees. I learned that, on average, these nest-site scouts devoted 37 minutes (range: 13–56 minutes) to the first inspection of the nest box, and that this first inspection comprised 10–30 forays inside the box. During each trip inside, the scouts walked around and made a few short flights. Most of these interior inspections lasted less than a minute. They were followed by time spent outside, when the scouts would hover around the box while facing it and often within two inches (5 centimeters) of it. I call this first visit, when a scout examines the potential dwelling place closely (both inside and outside), her "discovery visit." When it is completed, she returns to the swarm. If she has judged the site highly desirable, then she will advertise it with a long-lasting waggle dance, and she will visit it repeatedly, but her later visits become shorter than her early ones.

How does a scout bee get a sense of the roominess of a potential homesite? Is it simply a matter of going inside and looking around? The answer is no. I learned this when I returned to Appledore in the summer of 1976 and conducted an experiment with a special nest box that I had built (Fig. 3.5, top left). Its design was such that I could easily change its volume and its interior light level. When I darkened the box's interior by placing a light baffle inside its entrance hole, I could tell that the scouts were still able to sense its size. The scout bees "told" me this because whenever I shrank the space available in the darkened box (by inserting an inner lid) from 0.90 cubic feet (25 liters) to 0.18 cubic feet (5 liters), most of the scouts stopped visiting the box. They had lost interest in it. But an hour later, when I restored the space in the darkened box to 0.90 cubic feet (25 liters) by removing the inner lid, the number of scout bees visiting it rose. So, even when it was very dark inside the nest box, if it offered spacious living quarters, then the number of scouts visiting it grew, but if it offered cramped quarters, then the number of scouts visiting it shrank.

How is it possible for a nest-site scout to sense when a dark cavity is too small to be a good homesite? Might she develop a feel for its roominess (or lack thereof) based on walking around inside the cavity? I had seen that when a scout bee makes her first inspection of a potential home site, she devotes roughly 70 percent of her time to walking rapidly—at 2–3 inches (5–8 centimeters) per second—all around on the cavity's inner surfaces. This high-speed walking is interspersed with short "hop" flights lasting less than 1 second, during which the bee "jumps" from one point to another inside the cavity. I suspected that the rough landings of these brief interior flights were the sources of the bumping sounds that I had heard inside the bait hive I'd nailed to the pine tree behind my parents' house back in August 1974.

To test the role of walking in the process by which a scout bee senses a cavity's volume, I performed an experiment suggested by Professor Bert

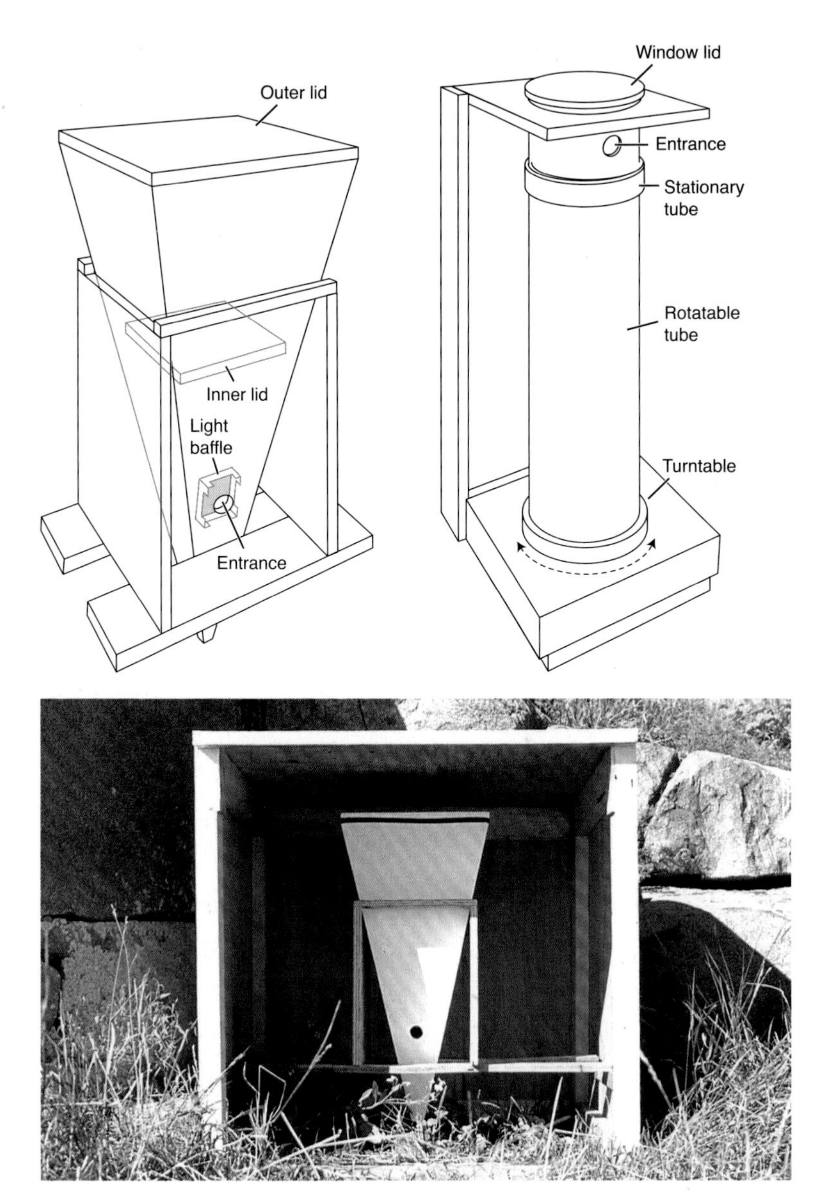

FIG. 3.5. Apparatus used to determine how a scout bee senses the volume of a cavity. *Top left*: Nest box in which the volume could be adjusted between 0.90 cubic feet/25 liters (without inner lid) vs. 0.18 cubic feet/5 liters (with inner lid). The light baffle enabled me to vary the light level inside the box. *Top right*: Apparatus in which the wall of the cylindrical cavity could be rotated to increase or decrease how much walking a scout bee needed to do to circumnavigate the cavity. *Bottom*: How the nest box shown in top left was presented to the bees, inside a lean-to shelter.

Hölldobler, my primary Ph.D. thesis advisor at Harvard: manipulate the amount of walking required of a scout bee to move all the way around inside a potential nest cavity that has a marginally acceptable volume, 14 liters (0.5 cubic feet). Would shortening (or lengthening) how long it takes the walking bee to circumnavigate the cavity cause her to sense that it is too small (or sufficiently large) to be a proper homesite? I brought a swarm to Appledore Island, and I set up a special nest cavity that consisted of a vertical cylinder with rotatable walls but a stationary entrance (Fig. 3.5, top right). I mounted this nest box inside a dark hut. Its entrance was connected to a hole in the side of the hut. Using this special nest box, I learned that if I shortened (or lengthened) how far the first scout bee had to walk to circumnavigate this cavity while making her inspections, then she would not (or would) recruit other scouts to my nest box. This showed that her assessment of this nest box's volume depended on her experience while walking around inside it. Evidently, a scout integrates the information acquired as she walks around inside a potential homesite to evaluate its roominess. I suspect that she uses the brightly lit entrance opening as a reference point to know when she has walked all the way around inside a cavity.

In 1950, Karl von Frisch wrote the following words in his little book titled *Bees: Their Vision, Chemical Senses, and Language*: "No competent scientist ought to believe these [discoveries] on first hearing." I agree. But Karl von Frisch's evidence is compelling, so I do believe his discoveries, just as I believe those of Martin Lindauer and many others who have studied the behaviors and cognitive abilities of worker honey bees. The experimental evidence is rock solid. Moreover, I have seen firsthand how a nest-site scout can make a multi-point inspection of a potential homesite, then form an assessment of its overall desirability, and finally express her sense of its goodness in the duration of the waggle dance that she performs to advertise the site. And, as you shall see in coming chapters, there is much, much more to her behavioral sophistication. Main take-home lesson for now: a worker honey bee is a remarkably canny creature.

CHAPTER 4

Choosing a Homesite

In spring or early summer, if a colony of honey bees has grown populous and has stocked its nest with brood and food, then almost without fail it will cast a swarm. When it does, the colony's queen and about three-quarters of its workers (some 12,000 bees) will tumble out of their home and form a cloud of swirling and buzzing bees. They will then assemble on a tree branch, mailbox, or whatever is nearby, and form a dense cluster with the mother queen safe inside. Left behind are a quarter or so of the colony's population of adult worker bees, a stock of several thousand immature worker bees, and a clutch of 4–10 immature queen bees being fed royal jelly. One of these daughter queens will replace the departed mother queen as the residual colony's egg-layer-in-chief. Also left behind is a larder of beeswax combs stocked with honey and pollen. It guarantees that none of the stay-at-home bees goes hungry, even though their group now has rather few foragers.

For a day or so after the first ("prime") swarm departs with the mother queen, the daughter queens remain motionless inside their special peanut-shaped cells. These hang from combs in the carefully warmed brood-nest region of the colony's home. Eventually, the daughter queens will complete their development and will begin to chew open their cells to get out. The first several to do so are likely to leave in another swarm (an "afterswarm"), but those that emerge later will stay at home and compete in deadly duels using their venomous, saber-like stings. The sole survivor of this queen-on-queen combat inherits a fully functional nest, the stores of pollen and honey it contains, and a sizable force of worker bees.

What happens to the bees that leave in a prime swarm or an afterswarm? Usually, they fly less than 300 feet (100 meters) from their home and then settle on a tree branch or the like. Here they form a beard-shaped cluster of bees that shelters the queen (if a prime swarm) or queens (if an afterswarm) (Fig. 4.1). Soon, a few of the worker bees will begin to perform waggle dances on the surface of their swarm's cluster. These are nest-site scouts, and their dances are announcements of the potential homesites they have discovered. As we shall see, this dancing lies at the heart of how the scouts choose a cozy new residence for their swarm.

Beekeepers have long known that worker bees perform dances on the surfaces of swarms, and some had suspected (correctly) that the dancing bees are nest-site scouts. But it was Martin Lindauer who first watched these dancing bees carefully and thereby studied closely the intricate process whereby the nest-site scouts in a swarm choose its new homesite. If you are a beekeeper, then I encourage you take a few minutes to watch the dances of nest-site scouts on the surface of a swarm the next time you collect a swarm. You will be rewarded with a glimpse of the most impressive form of democratic decision making there is outside the realm of human affairs.

Lindauer's studies of house hunting by honey bees began in the spring of 1949. That was when he found a swarm that had left one of the hives at the Zoological Institute of the University of Munich and then had settled on a bush nearby. Lindauer saw bees performing waggle dances on the swarm's surface, and he noticed that these dancing bees acted strangely: none carried loads of pollen, and none paused to unload nectar. Instead, they danced on and on without interruption. Furthermore, Lindauer saw that some of these dancing bees carried red brick dust on their bodies, while others were blackened with soot and smelled like chimney sweeps. He realized then that these bees had been poking around inside cavities

FIG. 4.1. A beautiful swarm of honey bees. It consisted of one queen bee, about 15,000 worker bees, and a few dozen drones.

and chimneys in the ruins of bombed buildings, and that they were scout bees engaged in finding a suitable home for their swarm.

At this point, the beekeeper from the Zoological Institute arrived and prepared to capture the swarm by shaking it into a hive. Lindauer asked her to let the swarm hang there so he could continue his observations, but she said, "That is out of the question. We need this colony!" The director of the Zoological Institute, Karl von Frisch, backed her up. Two years later, however, Lindauer got von Frisch's support to let the swarms from the Zoological Institute's hives hang undisturbed, so that he could watch them and study the behavior of their nest-site scouts (Fig. 4.2). During the summers of 1951, 1952, and 1953, Lindauer closely watched the activities of the scout bees on 19 swarms.

Lindauer was a pioneer in his studies of house-hunting honey bees. And as is often the case for people doing pioneering work, the tools available at the time were too crude to thoroughly investigate the uncharted scientific "territory" that lay before him. He once told me that when he did his studies his toolkit consisted of a chair, a wristwatch, a paint set, a compass, a notebook, and some pencils. He also explained that to make manageable his task of monitoring how many scout bees performed dances for each of the possible homesites they had discovered, he recorded only the number of *new* dancers for each site that appeared during a given time interval. He did this by putting a dot of paint on each dancing bee after he had watched her long enough to determine the direction and distance of the site she was advertising. He then ignored all subsequent dances performed by his marked bees. Even so, Lindauer's task was extremely difficult. Sometimes it must have felt almost overwhelming, such as when he saw several new (unmarked) scout bees performing dances indicating several different sites. Ideally, he would have recorded the *total number* of dancers—not just the number of *new* dancers—that advertised each site during each stage of the decision-making process. This would have revealed how the "popularity" of (i.e., the number of bees advertising) each potential homesite changed over time. But this was impossible in the 1950s, and so it stayed for more than 40 years.

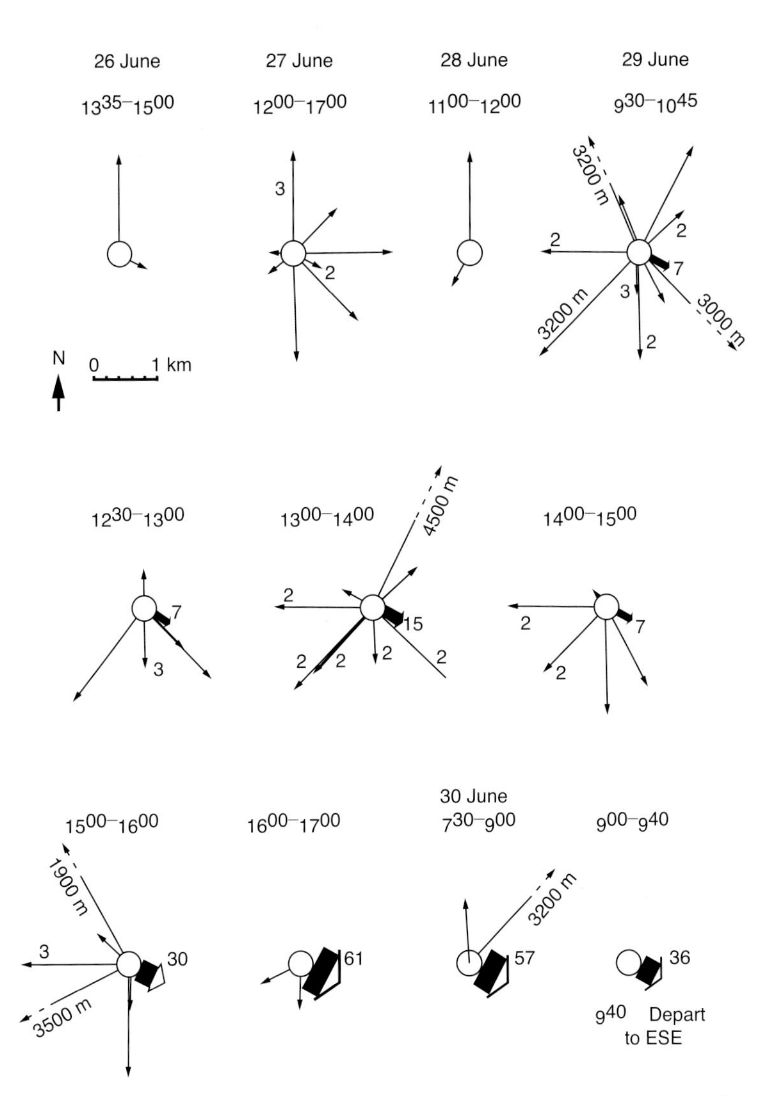

FIG. 4.2. The announcements made by the dancing scout bees in a swarm that Lindauer observed in June 1951. The numbers beside the arrows indicate the number of bees that started to dance for (i.e., advertise) a potential homesite within the time period shown. This figure shows that Lindauer's observations began shortly after the bees swarmed from their hive at 1:35 p.m. on June 26, and they ended when the bees flew to their chosen site at 9:40 a.m. on June 30. It shows, too, that more than 20 potential homesites were reported, but only a few were advertised by multiple bees, hence were important possibilities. For example, on June 29, between 9:30 a.m. and 4:00 p.m., a total of nine scout bees began to advertise a site located 1,500 meters (1,640 yards) to the west. Eventually, however, the scouts' interest in this site faded; Lindauer recorded no new dancers for this site after 4:00 p.m. Only one site, located 300 meters (330 yards) to the southeast, held the scouts' interest all day on June 29. Lindauer recorded more and more new dancers for the southeast site on this day, and between 4:00 and 5:00 p.m. the advertising of this site overwhelmed that of the other sites. The situation continued on the morning of June 30, with 93 out of the 95 new dancers advertising the southeast site. Finally, at 9:40 a.m., the swarm launched into flight, flew 300 meters (330 yards) to the southeast, and then moved into a nook in the wall of a bomb-damaged building.

In the summer of 1997, I was able to extend Lindauer's pathbreaking study of how nest-site scouts work together to choose their swarm's new homesite. I did so by tracking all the dances produced by the nest-site scouts on several swarms. This was something that I had been keen to do back in 1974, after I had watched a swarm of bees in my parents' backyard conduct its hunt for its future residence. Back then, though, it was still impossible to monitor closely all the individual bees performing waggle dances on a swarm. But by 1997, it had become possible! With funding from the National Science Foundation, I had purchased the high-resolution video recording and slow-motion playback equipment that would enable me to get complete records of the dances of the scout bees on swarms. Also by then, I had mastered the process of labeling thousands of worker bees for individual identification. Furthermore, in the summer of 1997, I had a research assistant—a Cornell undergraduate student, Susannah C. Buhrman—who was as determined as I was to figure out how the nest-site scouts work together. Across the months of June and July, Susannah and I prepared and studied three swarms, each of which consisted of 4,000 individually identifiable worker bees and their queen.

It took three days to prepare each swarm. It took this long because it involved repeating the following process about 40 times each day. First, we shook groups of about 10 worker bees (freshly removed from a hive) into small zip-lock plastic bags and then we put these bags of bees in a refrigerator. Once all the bees in a bag were in a chill coma, we poured them onto a block of "blue ice" to keep them chilled. Next, we glued a small plastic ID tag (bearing one of 500 color/number combinations) on the top of each bee's thorax. We also applied a dot of paint (one of 8 colors) on the top of each bee's abdomen. Finally, we poured each group of labeled bees into a shoe-box-size "swarm cage" which contained their queen. She was confined in a smaller cage about the size of a pack of chewing gum. The swarm cage was equipped with a feeder filled with rich sugar syrup, for we wanted our labeled worker bees to be like the worker bees in a swarm: stuffed with food. By repeating this individual

FIG. 4.3. Worker bees that have been labeled for individual identification. We did not know which bees in our swarms would be the nest-site scouts, and we did not want to disturb these scout bees during their work, so we labeled in advance every worker in our 4,000-bee swarms.

bee-labeling process 4,000 times, in marathon sessions from 8:00 a.m. to about 8:00 p.m. spread over three consecutive days, Susannah and I managed to label 4,000 worker bees, enough to make a small swarm (Fig. 4.3). We did all these things three times that summer, and so we managed to study three swarms composed of individually identifiable bees.

This project was a test of endurance. Getting every bee in a swarm labeled for individual ID was just the first step. Next, we needed to induce

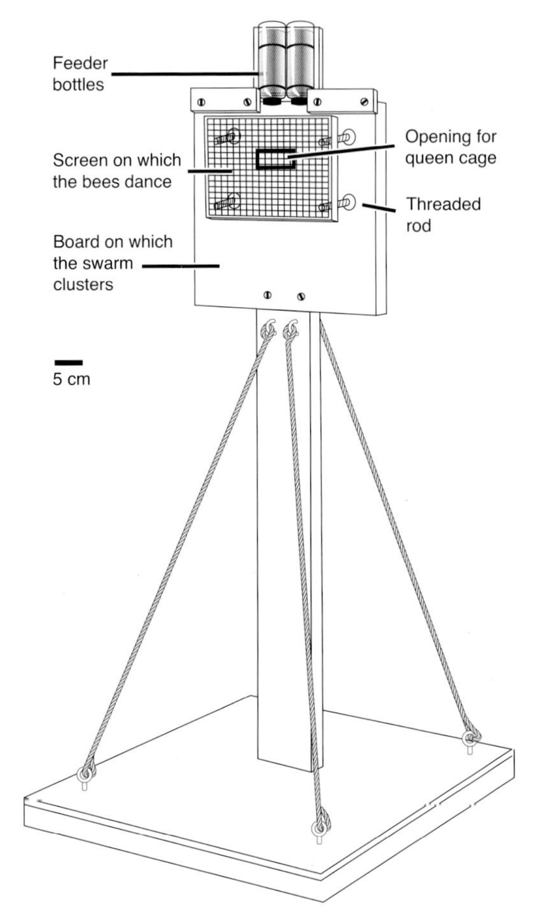

FIG. 4.4. The swarm board used for video recording the dances of the nest-site scouts.

each swarm (one at a time) to cluster on the side of a vertical board (the "swarm board," Figure 4.4). This way, the swarm's cluster would have just one side exposed to daylight, which meant that it would have just one surface on which the nest-site scouts would perform their dances. It was easy to get a swarm to cluster where we wanted; we simply fastened the cage holding the queen to one side of the swarm board. The worker bees then formed a cluster over her. Once the bees had done so, we positioned a wire-mesh screen beside the cluster so that the bees in its outer layer were standing (or walking) on this screen. This provided the nest-site scouts with a perfectly flat, and vertical, surface on which to perform their dances. Now we were ready to watch, and video record, all the waggle dances of the nest-site scouts (each one individually recognizable) throughout their process of choosing their swarm's new residence. As usual, the nest-site scouts performed their waggle dances on the surface of their swarm's cluster, not inside it, so this setup worked well. Once these scouts had completed their deliberations, they stimulated everyone else in the swarm to warm their flight muscles to prepare for takeoff, and then they triggered the entire swarm to launch into flight (see Chapters 6 and 7). These swarms could not fly away, however, because each swarm's queen remained behind, confined in her cage on the swarm board. Eventually, each swarm returned and reassembled around its queen. We then put all the bees in a hive. So, in the end, each swarm did "find" a good home.

The hard work was over for the scout bees, but it was just starting for Susannah and me. For each swarm, we needed to determine, by reading the scout bees' dances, the location of the potential homesite that each scout bee had advertised each time she performed a dance on the surface of the swarm cluster. This required analyzing every dance that appeared throughout the 12–16 hours of video recording per swarm. Doing so yielded something that I still find pretty amazing: a detailed record of the scout bees' debate on each of our three swarms. (*My dictionary defines a debate as "a formal discussion of some question of public interest in a legislative or other assembly," so I feel that the word "debate" applies to the group decision making that is done by the nest-site scouts in a honey bee swarm.*) In other words, for each of the hundreds of dances performed on each swarm, we knew the ID of the nest-site scout that performed the dance, the location of the potential homesite that she had advertised with her dance, and even the level of her enthusiasm about her site. A nest-site scout expresses her enthusiasm for a site by the persistence of her dancing. Some of the scouts on our swarms advertised their sites with feeble, stop-and-start dances that petered out after just 5–10 circuits, while others promoted their sites by dancing steadily and strongly for

100 or more circuits. It was wonderful to see a nest-site scout land on the swarm's surface and immediately begin dancing so excitedly that, if she were a person, one might say, "She could not get the words out of her mouth fast enough!"

The most intriguing decision process that Susannah and I watched is the one we recorded in our final (third) swarm, for in this one there unfolded a fierce debate between two groups of scout bees (Fig. 4.5). We set up this swarm on the afternoon of 19 July 1997, but it was not until late morning the next day that the scouts started to report possible homesites. Between 11:00 a.m. and 1:00 p.m., six sites (A–F) were announced, and one (site A) gained a lead with eight scouts performing dances to promote it. Between 1:00 and 5:00 p.m., three more possible homesites (G, H, and I) were reported, so at 5:00 p.m. it looked like the contest among the dancer groups was still wide open. However, the competition tightened up a lot during the last two hours of the scout bees' debating on this day. Between 5:00 and 7:00 p.m., only sites B and G were advertised by multiple dancing bees. It was impossible to tell, though, which site would become the winner. Susannah and I made bets on whether the scouts from site B (my pick) or site G (Susannah's pick) would prevail.

We did not want to miss seeing how this discussion would play out, so we arrived at the lab shortly after sunrise the next day, got our video equipment set up, and then waited for the air to warm enough for the nest-site scouts to resume their debate. Which group of scout bees would prevail? And whose bet would prove correct? For the first two hours, 7:00 to 9:00 a.m., the outcome remained uncertain. Both sites were advertised by about a dozen bees producing persistent dances, but those for the southwest site (G) lasted a bit longer than those for the south site (B): 20 vs. 15 circuits per dance, respectively. This difference was enough to break the tie. Over the next two hours the scouts advertising site G built a strong lead over those promoting site B: 32 vs. 17 dancers, respectively. Rain started at 11:54 a.m. and continued through the afternoon and evening, so the outcome of this debate remained uncertain until the following morning. The scouts resumed dancing a little after 9:00 a.m.

and now, somehow, they were unanimous! Out of 73 bees, 73 danced for site G in the southwest. A few minutes before noon, the swarm launched into flight and flew off to the southwest. We bid them farewell, and soon headed off to Ben and Jerry's to settle our bet.

It was fascinating to watch the dance-offs among the scout bees on our three swarms. But, as is often the case in research, we did not appreciate fully what we had seen until we prepared figures that displayed visually what had happened, i.e., how the "contests" among the scout bees in each of our three swarms had played out. Building the three summary figures for our three swarms required watching slowly and painstakingly more than 48 hours of video recordings. We had to extract from these recordings three pieces of information about each scout bee that had performed waggle dances: (1) when she performed her dances, (2) how long each of her dances lasted, and (3) what site she advertised in each of her dances. Fortunately, the numbers of bees that performed dances advertising homesites were fairly modest: "only" 73, 47, and 149 bees per swarm.

Even so, it took more than 500 person-hours to extract from the video recordings a complete record of the dancing by all the nest-site scouts in our three swarms. It took this long because it required examining one by one the 1,348 dances that the nest-site scouts performed. And for each dance, we needed to determine four things: (1) which bee produced it, (2) when it was performed, (3) how long it lasted, and (4) what site it advertised. We persisted with this work because we knew that getting a dance-by-dance look at what happened in our three swarms was the only way we would ever solve the mystery of how the nest-site scouts conduct their collective decision making.

Eventually, we had before us what we needed for each swarm: a complete register of which individuals had worked as nest-site scouts in a swarm, and for each of these bees, *when* she performed each of her dances, *which site* she advertised in each dance, and *how many* circuits she produced in each dance. Next, we made a set of diagrams for each swarm (e.g., Figure 4.5 for Swarm 3) that shows when each potential

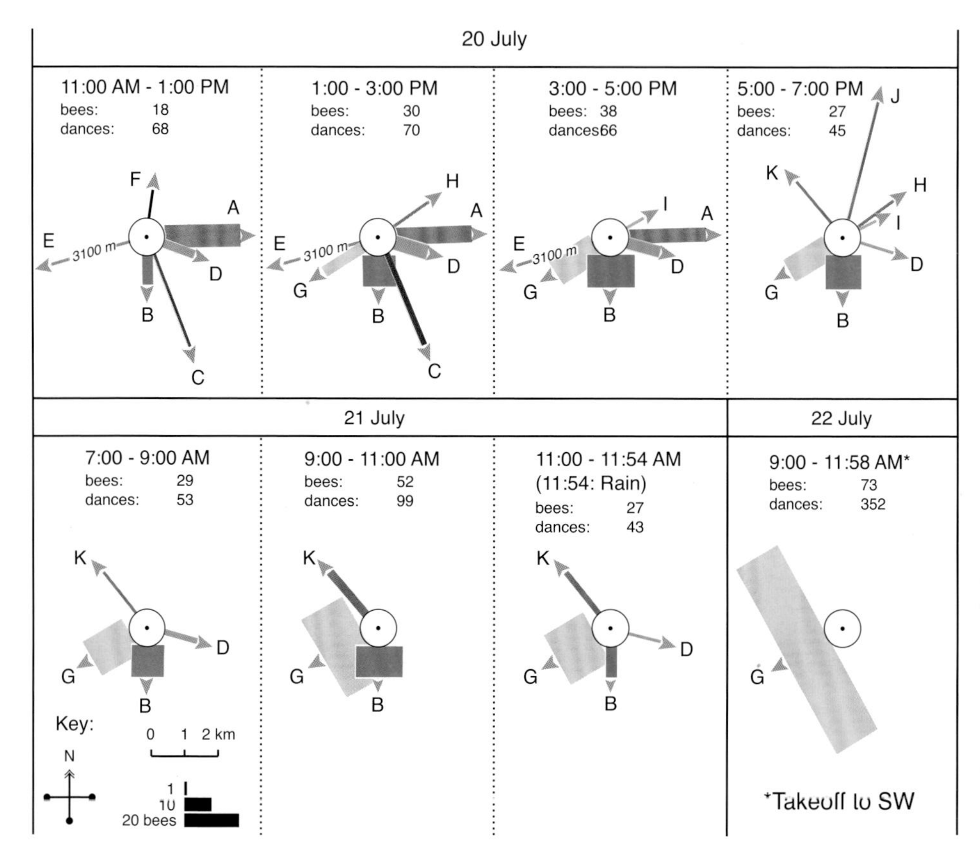

FIG. 4.5. The pattern of dances produced by nest-site scouts on the swarm observed from 20–22 July 1997. This figure shows that the swarm's scout bees reported 11 possible homesites (A–K) and these bees eventually reached a consensus decision in favor of site G. This consensus was achieved, however, only after a prolonged competition between the scouts from sites B and G.

homesite was first advertised, where it was located, and how the number of scouts that advertised it changed over time. Our diagrams were inspired by Martin Lindauer's (see Fig. 4.2), but they differed from his by showing how many scouts *in total* advertised each site within a certain time interval, not just how many *new scouts* did so. Thanks to technological advances since the 1950s, we were able to produce a fuller, and more telling, picture of how the scouts' interest rose and fell for each potential homesite.

What did we learn about how the nest-site scouts perform their collective decision making? *First*, we learned that the scout bees' deliberations started slowly, with an information accumulation phase during which the scouts put a dozen or so alternatives "on the table" for discussion. In each of the three swarms that Susannah and I watched, the number of sites considered was 13, 5, and 11. *Second*, we learned that the debates in our three swarms ended with all or nearly all of the dancing scout bees advocating just one site, that is, showing a consensus. *Third*, we learned that the decision-making process in each of our swarms involved many dozens of scout bees; hence it was a highly distributed process. As mentioned already, Susannah and I observed 73, 47, and 149 bees (average 90 bees, or about 3 percent of the swarm) performing dances in the three swarms that we studied. These counts, however, probably underestimate the typical number of dancing bees in a swarm; we used small swarms—with only 3,252, 2,357, and 3,649 bees—to keep doable our task of labeling every bee in our swarms for individual identification. Natural swarms contain, on average, about 12,000 bees—four times the size of ours—so it is likely that a typical swarm will have at least 300 to 400 scout bees involved in the choice of its new home.

I think it makes sense for a honey bee swarm to have hundreds of its members engaged in choosing its new residence. This is a decision that must be made *accurately*, because a colony needs a sufficiently roomy, snug, and secure homesite if it is to survive. Furthermore, it is a decision that must be made *speedily*, because a swarm almost always hangs from a tree branch or the like, so it is exposed to the weather. Dispatching hundreds of nest-site scouts to find several possible dwelling places, and then holding a "friendly" debate among these scout bees to identify which one is best, looks to me like a very smart way for a swarm to choose its future living quarters.

CHAPTER 5

Consensus or Quorum?

Biologists usually study the common order of things. Sometimes, though, it is good to pay attention to an anomaly, that is, a deviation from what one commonly sees, because it might provide a clue that helps solve a mystery about the natural world. Charles Darwin, for example, wrote in his famous book of 1859, *The Origin of Species*, "There is no greater anomaly in nature than a bird that cannot fly; yet there are several in this state." Darwin was referring to several species of nearly wingless birds (e.g., the kiwi, *Apteryx mantelli*, in New Zealand) that live on oceanic islands without predators, and he did so because he saw that these birds help us understand how natural selection works. Where there are no predators on the members of a bird species, the benefits of wings do not outweigh their costs, so natural selection reduces the wings in this species over successive generations. This morphological anomaly helped Darwin show the role of natural selection in shaping how organisms are built. In this chapter, we will see how a behavioral anomaly in two honey bee swarms helped me and a friend, Kirk Visscher, who is a fellow honey bee biologist, figure out how a swarm of honey bees knows when to make the shift from quietly hanging from a tree branch to quickly flying to its new homesite.

In the previous chapter, we saw how the nest-site scouts in a swarm advertise the sites they have found by performing waggle dances, and that soon after all (or nearly all) the scouts are advertising the same site, the swarm takes off and flies to the agreed-upon homesite. This pattern of events suggests that a swarm's takeoff is triggered by the scout bees reaching a consensus about which site will be their swarm's new dwelling place. This is how it looks, but this is *not* how it works. We know this because occasionally one sees an anomaly: a swarm takes off when its scout bees are vigorously performing waggle dances for two sites. Then the airborne bees separate into two groups that move off in different directions, but eventually the two groups rejoin and resettle because there is just one queen. So how does this anomaly arise? Solving this mystery revealed that what triggers a swarm to launch into flight is not the scouts *reaching a consensus* at the swarm cluster, and is instead the scouts *building a quorum*—a sufficient number of scout bees—at the new homesite.

The first observation of a swarm of honey bees launching into flight when its scouts had not formed a consensus was made by Martin Lindauer on 22 June 1952. On this day, he was watching a swarm that had settled on the balcony over the front entrance of the Chemistry Institute of the University of Munich, which was less than 600 feet (200 meters) from Karl von Frisch's apiary at the Zoological Institute. Lindauer dubbed this swarm the "Balcony Swarm." He watched it closely and saw that its scout bees got into a competition: one group of dancers favored a site about 2,000 feet (600 meters) to the northwest and a second group favored a site about 2,600 feet (800 meters) to the southwest (Fig. 5.1). For five hours (11:00 a.m. to 4:00 p.m.) neither group managed to gain a decisive lead. Nevertheless, at 4:10 p.m. the swarm lifted off and then did something that Lindauer could scarcely believe even though he was watching it happen. Here are his words (translated from German): "The swarm . . . sought to divide itself. The one half wanted to fly to the northwest, the other to the [southwest]. Apparently, each group of scouting bees wanted to abduct the swarm to the nesting place of its own choice." And each group partly succeeded, for half of the airborne bees started moving southwest (toward Munich's main railway station) while the other half began moving northwest (toward a heavily bombed residential area). But neither group managed to continue flying to its destination,

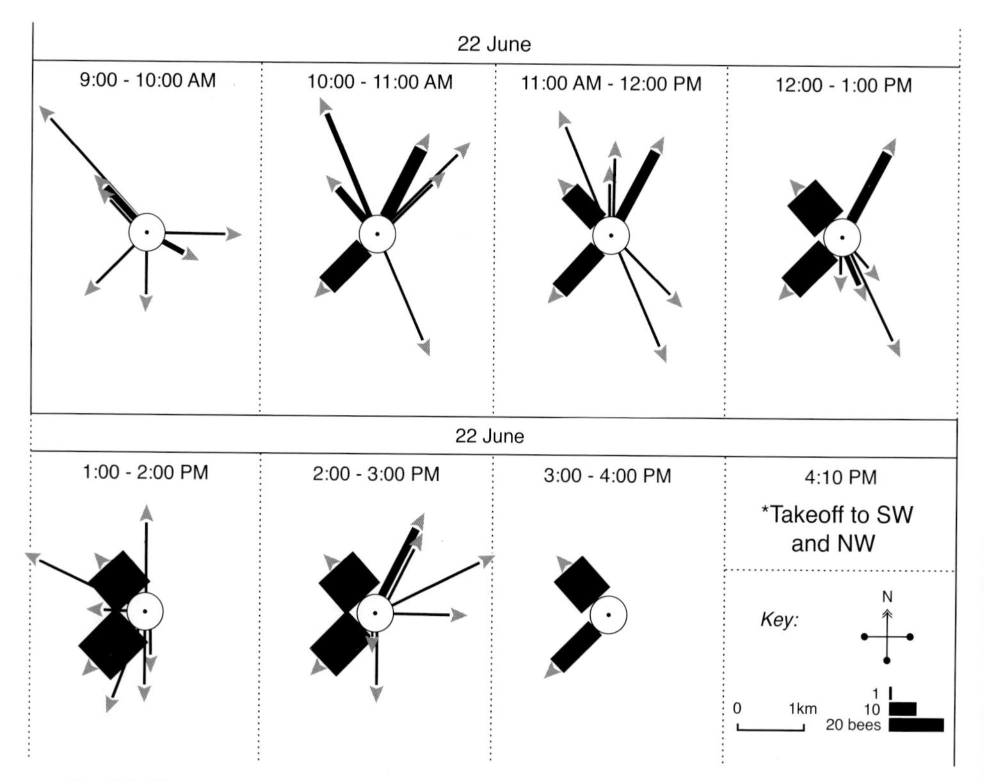

FIG. 5.1. The announcements made by the dancing scout bees in the "Balcony swarm" that Martin Lindauer observed in June 1952. Formatting is the same as in Fig. 4.2. This swarm's scouts never reached a consensus, and yet at 4:10 p.m. on 22 June the swarm lifted off and tried to move to two new homesites, one about 2,000 feet (600 meters) to the northwest and the other about 2,600 feet (800 meters) to the southwest.

probably because neither group included the queen, and eventually the two clouds of swirling bees reunited in the air where they had started. The bees then resettled on the balcony. Sadly, the swarm's queen had become lost during the aerial tug-of-war. Over the next several hours Lindauer watched this swarm's cluster dissolve as the bees drifted back to their hive in the apiary at the Zoological Institute.

The next chapter in this story came 50 years later, on 7 July 2002. This is when Kirk Visscher—a professor in the Department of Entomology at the University of California, Riverside—and I observed an anomaly similar to what Martin Lindauer had reported. Kirk and I were together on Appledore Island, conducting a study to determine how the nest-site scouts know when they have completed their decision making about the future homesite. We had wondered whether the nest-site scouts do so by sensing *a consensus* (an agreement of dancing scouts) at the swarm cluster, or by sensing *a quorum* (a sufficient number of scouts) at one of the potential nest sites. We knew that the nest-site scouts usually reach a consensus in their dances before the swarm moves to its new home (as shown in Figures 4.2 and 4.5). But we also knew that Lindauer had seen a swarm launch into flight even though its nest-site scouts had *not* reached a consensus in their dances. It was clear that this matter deserved a closer look. We decided to investigate this mystery in a setting where we could control the housing options of a swarm, figuring that this would enable us to monitor closely both the dances of nest-site scouts on a swarm and their activities at prospective homesites. Appledore Island was the perfect place for our study.

It felt good to return to Appledore Island. Much remained in 2002 as it had been when I first came out to this island in 1975. Herring gulls still staunchly defended their territories, swooping down from behind and thwacking our heads with their feet if we came near their nests. Waves still seemed to crash in slow motion over the deadly ledges around Duck Island, a mile to the north. Thriving bushes of poison ivy remained a danger along many of the trails on Appledore. And sunlight still shone as fiercely as ever, especially atop the island, where a porch on the old Coast Guard Life Saving Station provided a perfect spot for setting up (one by one) our swarms (Fig. 5.2). One thing was different, though: my lobster fisherman friend, Rodney Sullivan, had retired and had sold his house to some folks from Massachusetts. They were using it as a summer home. So, I didn't go over to the Sullivan place and "fix" its chimney before we set up our first swarm. Kirk and I would rely instead on monitoring the dances of the nest-site scouts and censoring (caging) any scout bee that reported a potential homesite other than our nest boxes.

The research plan that Kirk and I had prepared was straightforward. We would present four swarms, one at a time, with two identical nest

FIG. 5.2. The main building of the former Coast Guard Life Saving Station on Appledore Island, built in 1910. The perspective is looking east. The small porch on the right is where we set up our swarms and then watched and video recorded the activities of their nest-site scouts.

boxes that offered the nest-site scouts two superb homesites. Our aim was to induce a strong debate among the scouts in each of our four swarms, and then see if any of these swarms would take off to move to a nest box *before* their scouts had reached an agreement in their dancing. If at least one swarm did so, then this would confirm what Lindauer had observed with his Balcony swarm: the nest-site scouts did not rely on sensing a dance consensus to know when to excite their swarm to launch into flight.

We conducted four trials of our experiment, each with a different swarm. For each trial, we positioned our test swarm on a porch of the former Life Saving Station, and we set out two nest boxes on the east side of the island. Both boxes were positioned 820 feet (250 meters) from the swarm, but they were in different directions (approximately NNE and

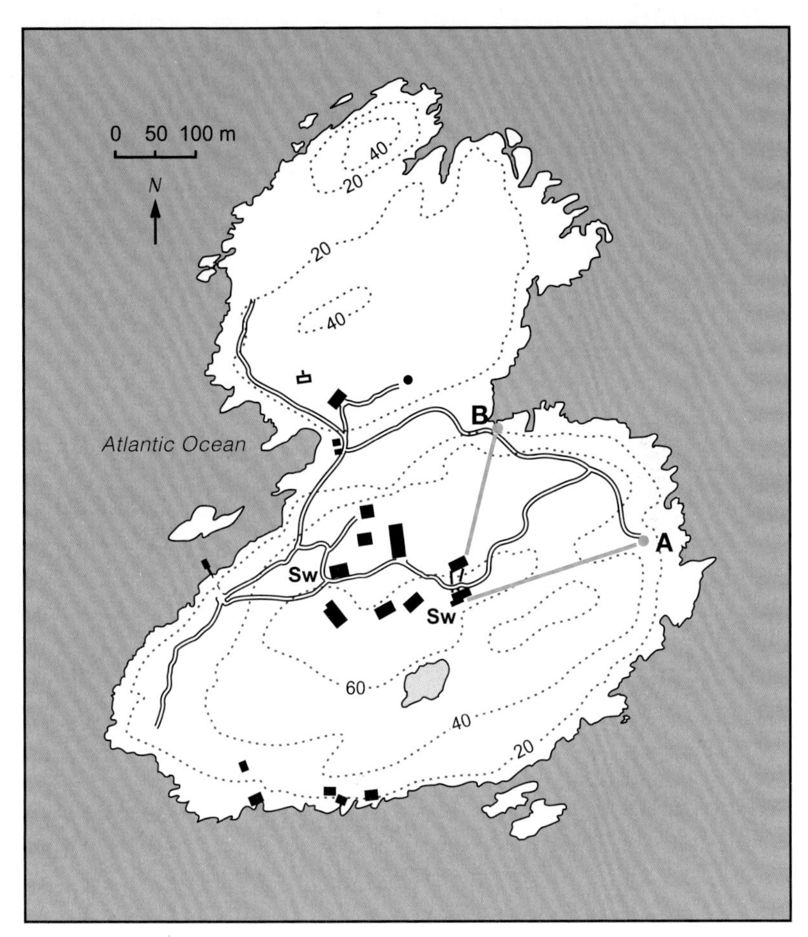

FIG. 5.3. Layout of the experiments on Appledore Island in 2002 and 2003. In 2002, we set out one nest box at both sites A and B. In 2003, we set out one nest box at site A and five nest boxes at site B.

ENE) from the swarm. So they were about 740 feet (220 meters) apart (Fig. 5.3). We wanted to census the scout bees both inside and outside each nest box, so we mounted each one against a window on the side of a lightproof hut, just as I had done 27 years earlier (see Fig. 3.4) to see how scout bees perform their interior inspections of a potential homesite.

This plan worked well. We found that our swarms would discover both nest boxes at about the same time, would develop a balanced debate over these two attractive homesites, and sometimes would take off when both sites were being advertised by scout bees performing waggle dances on the swarm cluster. This showed that the scouts did not need to reach a full consensus to make a decision. But what was most telling was the phenomenon that we watched unfold on 7 July 2002. By 11:30 a.m., the number of scout bees had reached a high level (more than 50 bees total, inside and outside) *at both boxes*. Then, at 12:04 p.m., when both boxes were being advertised strongly by scout bees feverishly performing dances, the swarm took off and then, to our surprise . . . it split itself! Separate clouds of bees formed on the north and east sides of the Coast Guard building, and at 12:09 each group began to move slowly in the direction of "its" nest box. Both groups, however, traveled only about 130 feet (40 meters) toward their destinations and then stopped. We supposed that the queen was missing from both groups, and we feared that she was lost. But at 12:15, I spotted her at the Coast Guard building, walking on the floor of the porch where we had set up the swarm. It seems that she had failed to fly off with either group of worker bees. I picked her up, put her in a little cage, and fastened it to the swarm board. Soon both groups of worker bees began to return and resettle around her.

Kirk and I were delighted by this swarm's behavior, for it showed us that a consensus among the nest-site scouts dancing on the swarm is *not necessary* for a swarm to initiate its move to a new home. It also confirmed the anomaly that Lindauer had seen with his Balcony swarm 50 years earlier: a swarm launching into flight when the dances of its scout bees were split between two sites, and then splitting itself into two groups of bees that moved off in separate directions. When I shared the news of our experimental results with Martin Lindauer, then 84 years old and living in a village outside Munich, he was delighted, for now his "strange" story of an airborne swarm splitting into two separate clouds of bees had been confirmed.

So, a consensus among dancing nest-site scouts *is not* necessary for a swarm to initiate its move. But what *is* necessary? What tells the nest-site scouts that the time has come to trigger their swarm's takeoff? It seemed to Kirk and me that it is probably the presence of a sufficient number (i.e., a quorum) of scout bees at a potential homesite that tells these bees that it is time for them to excite their swarm mates to take off and move to their new residence. We suspected that the scouts rely on quorum sensing because in the course of conducting the four trials of our experiment in 2002, we had seen that each swarm had started its flight preparations—that is, the nest-site scouts had begun to make shrill piping signals at the swarm (see Chapter 6)—when the number of scouts at one of the nest boxes had risen to 20–30 bees. Usually, this was 10–15 bees inside the box and another 10–15 outside it.

In June 2003, Kirk and I returned to Appledore Island with the aim of testing the quorum sensing hypothesis. Our plan was to test a critical prediction of this hypothesis: delaying the formation of a quorum of scout bees at a swarm's chosen nest site, while leaving the rest of the decision-making process undisturbed, should delay the start of piping by the nest-site scouts and thus the takeoff of the swarm. We had a simple way to delay quorum formation: place five perfectly good nest boxes close together at one site on Appledore Island (Fig. 5.4). This was site B, shown in Figure 5.3. By providing five boxes at one site, we caused the scouts visiting this site to be dispersed among the boxes. We also performed a control trial with each swarm, either before or after the experimental trial. For it, we placed just one nest box at our other site on Appledore: site A in Figure 5.3. In both the treatment trial and the control trial, we saw how long it took a swarm's scout bees, once they had discovered the site with the nest boxes (or nest box), to trigger their swarm to fly to the site.

We tested four swarms, and with each one we observed that the number of scout bees built up rapidly at the lone nest box in the one-nest-box trial, but that their numbers built up slowly at each nest box in the five-nest-box trials. This was because in the latter trials, the scout bees distributed themselves very evenly among the five nest boxes. We also observed that each swarm was much faster at reaching a decision in its one-nest-box trial (average: 196 minutes) than in its five-nest-box trial

FIG. 5.4. The cluster of five nest boxes (each one is sheltered within an orange lean-to) along a path on the eastern side of Appledore Island. Every 15 minutes, a count was made of the nest-site scouts visible outside each nest box.

(average: 442 minutes). This experiment produced strong support for the quorum-sensing hypothesis. Exactly *how* the scout bees sense a quorum, however, remains a mystery. Perhaps the scout bees do so visually (as we did), by seeing a crowd of other scout bees flying slowly around at a site. Or maybe they do so by noticing that they bump into lots of other scout bees when they go inside a potential nest site to inspect its interior.

Why do nest-site scouts rely on quorum sensing rather than consensus sensing to know when they have reached a decision? I think that one reason they do so is because a swarm needs to have a sizable group of bees that have visited the new homesite and so are able to guide everybody to this site. Encountering numerous fellow scouts at a site is a reliable indication that this need has been met. Quorum sensing may also be a way to help ensure that a swarm makes a good decision. The requirement of a quorum of at least 20–30 scout bees present simultaneously at a future homesite certainly helps to ensure that a swarm chooses wisely. The presence of this many bees means that numerous scouts have examined the site and judged it desirable. But why is 20–30 bees the critical number of bees rather than, say, 10–15 bees or 40–60 bees? I suspect that the quorum size of 20–30 scout bees is a part of the honey bee's house-hunting process that has been tuned by natural selection to give a swarm an optimal balance between speed (favored by a small quorum) and accuracy (favored by a large quorum) in choosing its homesite. A swarm must choose quickly, lest it be caught outdoors in bad weather. A swarm must also choose carefully, so it will be protected well from foul weather and fierce attackers in its new home.

If you are a beekeeper who likes to collect swarms to have more colonies, then it is good to know that the scout bees in a swarm rely on quorum sensing at a homesite rather than consensus sensing at the swarm, to know when they have finished choosing their new home. Knowing this warns you that a swarm can launch into flight even if you see, on the surface of a clustered swarm, scout bees performing waggle dances that advertise multiple home sites. Fortunately, there does exist *a very reliable indicator that a swarm is about to launch into flight*: the sound of nest-site scouts within a swarm's cluster producing a high-pitched sound called "worker piping." This handy sound signal is the subject of the next chapter.

CHAPTER 6

Piping Hot Bees

One of the joys of the biology of honey bees is the diversity of their behavioral signals. Besides the famous waggle dance, there are the queen and worker piping signals, the buzz-run signal, the tremble dance, the shaking signal, the beep signal, the grooming invitation dance, and still more. Each one sends a specific message from the bee producing the signal to the bee (or bees) receiving it. For a behavioral biologist like me, each signal poses a bundle of mysteries. How is it produced? How is it detected (even in the darkness of the hive)? And what is its role in the functioning of a colony? We will examine most of these behavioral (i.e., non-chemical) signals in this book, and we will start now with the piping signal of worker bees. This is an acoustical/mechanical signal that consists of a pulse of a distinctive, high-pitched sound. Each pulse of this sound lasts about one second and has an upward sweep in its pitch. The piping signals made by workers resemble those made by queens. Even if you have never heard the sound of a piping queen, you can get a sense of the sound of a piping worker by imagining a much softer version of the sound made by a Formula One racing car as it makes a full-throttle acceleration.

I first heard worker bees produce piping signals back in the summer of 1970. I had just graduated from Ithaca High School and was working for Doc Morse (Fig. 6.1) as a helper at the Dyce Laboratory for Honey Bee Studies. My job responsibilities included mowing the lab's lawn,

FIG. 6.1. Professor Roger A. Morse (Doc) in September 1975, demonstrating to Cornell students how to open a bee hive, handle the frames of comb, and locate the queen. Helping him is one of his graduate students at the time, David DeJong.

sweeping its floors, painting hives, running errands, helping with the beekeeping, and assisting graduate students and visiting scientists with their projects. I enjoyed all these tasks, but my favorite experience that summer came when I worked as Doc's eyes and ears in an experiment that needed somebody to monitor two honey bee swarms from sunrise to sunset for five consecutive days.

The aim of this experiment was to see how closely the workers in an airborne swarm monitor the presence of their queen throughout their flight to the new homesite. To set up this experiment, Doc collected two swarms and found the queen in each. He then put each swarm's queen in a small cage and wired this cage to the horizontal bar of a wooden cross about five feet (1.5 meters) tall. The cross functioned as a swarm mount (i.e., an artificial tree branch). Next, he let the worker bees in each swarm gather around their caged queen in a beard-like cluster. Doc's two crosses stood about 120 feet (40 meters) apart, on opposite sides of a small apple orchard. This enabled the two swarms to function independently, and to be monitored by just one person. The study site was an abandoned farmstead on top of Mount Pleasant, about 4 miles (6.4 kilometers) from Dyce Lab. Doc chose this open spot because he wanted me to see how far a queenless swarm would fly off before it returned to its caged queen.

My assignment was to watch Doc's swarms from 7:00 a.m. to 7:00 p.m. on five consecutive days, and to record when each one launched into flight and started moving to the homesite that its scout bees had selected. Also, I was to record how far each airborne swarm moved off, and when it returned and reassembled around its caged queen. The top of Mount Pleasant is fairly level and has good soils, so it is mostly covered with fields, but its sides are steep and are covered with forests (Fig. 6.2). Doc was confident that the scout bees in his swarms would find attractive homesites in the hollow trees in these woods. He was also confident that his swarms would attempt to move to the sites chosen by their nest-site scouts, but would fail to do so because their queens would be trapped in cages on his wooden crosses. Doc knew that the workers in a swarm will not abandon their queen.

FIG. 6.2. The open landscape atop Mount Pleasant, where I first watched and listened to swarms as they prepared to move to the homesites their nest-site scouts had chosen. The perspective is looking west from the highest point on this hill, which is 1,780 feet (540 meters) above sea level. The white building in the distance is an observatory of the Cornell Astronomical Society.

Over the five days that I watched Doc's swarms, I saw each one make several attempts to move to a tree cavity, and I recorded the times of each swarm's departures from and returns to its caged queen. The bees in these swarms had enough fuel to make repeated attempts because there was a feeder bottle filled with sugar syrup wired to each swarm's cross. Whenever a swarm launched into flight and attempted to fly to its new homesite, the cloud of airborne worker bees moved off slowly and did not go far, at most about 250 feet (75 meters). Meanwhile, the queen dashed around inside her cage, no doubt sensing that she should be flying away with her entourage of workers.

Thinking back to this project, I am pretty sure that Doc wanted the data that I collected on the departure and return times of his swarms so he could see how quickly the workers in an airborne swarm can detect and respond to the absence of their queen. He knew that each swarm would start to move toward its future homesite, but then would sense that its queen was missing so would come back and reassemble around her. On average, these swarms were away from their queens for about

10 minutes. In nature, queens do sometimes drop out of airborne swarms to rest, so Doc's experiment presented the worker bees in his swarms with a realistic problem. And the experiment's results showed clearly that worker honey bees are good at monitoring their queen's presence and are quick to respond to her absence. I will admit, though, that I never asked Doc about the purpose of this study. I was only a few weeks into the job and I suspect that my frame of mind was akin to what Alfred, Lord Tennyson described in his poem *The Charge of the Light Brigade*: "Theirs not to reason why, / Theirs but to do and die."

What I remember most clearly from helping Doc with this experiment was what I heard whenever I put an ear close to a swarm during the last hour or so before it launched into flight: a cacophony of high-pitched sounds. At first, they were intermittent, so they were distinct. Each one lasted about a second and had an upward sweep in its pitch. They reminded me of the piping sounds made by queen honey bees, so I thought of them as worker piping sounds. Eventually, these shrill sounds became numerous and loud, whereupon the cluster of bees disintegrated and made a cloud of bees swirling just above my head. It was clear that some worker bees within the cluster had been urgently sending a high-pitched sound signal to their fellow swarm bees. But what the message of this signal was, and how it was sent and received, remained mysteries to me for nearly 30 years.

The solutions to these mysteries emerged in stages. The first came in 1981, when Bernd Heinrich, a gifted insect physiologist (then at the University of California at Berkeley), studied the mechanisms of thermoregulation in honey bee swarms. Bernd had pioneered the study of body temperature control in insects, so he started his work with honey bee swarms with a broad knowledge of insect thermoregulation. He knew that two previous studies with swarms had revealed that the core temperature of a swarm cluster is about 95°F (35°C), hence the same as in a colony's brood nest. Bernd also knew that an individual worker bee can produce heat by shivering—isometrically contracting the two sets of flight muscles in her thorax—and that she needs to warm up her flight muscles to at least 95°F (35°C) to produce a wingbeat frequency that is sufficiently high to support flight (about 250 wingbeats per second). Furthermore, Bernd knew that before leaving their home, swarming bees fill their crops ("honey stomachs") with honey. This means that a swarm leaves home well supplied with fuel for keeping itself warm, for powering the flights of its nest-site scouts, and for secreting the beeswax it needs for comb building after it has moved into its new homesite. But neither Bernd nor anyone else knew about the temperature profiles inside a swarm's cluster. So, learning what these profiles are, how they are produced, and how they change as the ambient temperature rises and falls, were the aims of the study he made in the early 1980s.

Bernd discovered many marvelous things about temperature regulation in swarms, all of which are key to understanding how a beard of bees hanging from a tree branch prepares to fly to its new homesite. First, he found that the bees in a swarm precisely control the temperature in the core of their cluster; it stays at 93–97°F (34–36°C) regardless of the ambient temperature. He also found that the bees in a swarm allow the temperature in the outer layer of their cluster (its "mantle") to vary with the ambient temperature, but they keep its temperature above 63°F (17°C) even if the temperature of the air around the swarm cluster falls to freezing (32°F/0°C). This means that the outermost bees keep themselves warm enough to stay active. It also means that before a swarm can take off to fly to its new living quarters, the mantle bees must work (i.e., shiver) to warm their flight muscles to the flight-ready temperature of 95°F (35°C). And not just in theory. When Bernd made continuous recordings of the temperatures at various locations in a swarm's cluster, starting when the swarm settled and ending when it took off to fly to its new home, he found that during the last hour or so before takeoff, the temperature of the cluster's *mantle* did indeed rise to match the 95°F/35°C of its *core*. Impressive!

I was fascinated by what Bernd Heinrich had learned, and in 2002 I traveled to Germany to have an even closer look at the process of preflight warm-up by the worker bees in swarms. I did so using an infrared video camera. In 1989, my Ph.D. thesis advisor and friend, Professor Bert Hölldobler, had moved from Harvard University to the University of Würzburg and had become Chair of the university's Institute of Behavioral Physiology and Sociobiology. One of the many marvelous pieces of scientific apparatus that Bert purchased for the institute was an infrared video camera. This kind of camera enables one to measure the surface temperatures of multiple objects (such as worker bees) simultaneously and without disturbing them. An object's color in the video image indicates its surface temperature. I teamed up with two Ph.D. students, Brigitte Bujok and Marco Kleinhenz, and their advisor, Professor Jürgen Tautz, and we recorded the temperatures of the bees in the outermost layer (mantle) of two swarms. In each case, we started recording the temperatures of the surface-layer bees when the swarm bees formed their cluster, and we continued our recording until the swarm launched into flight. Our goal was simple: learn as much as possible about how the mantle bees in a swarm warm their flight muscles before their swarm takes off.

This work proceeded smoothly. Over a two-week period, we recorded the temperatures of the mantle (outer layer) bees within a 4 × 4-inch (10 × 10-centimeter) area on two swarms, starting when each one formed its cluster and ending when each one took off to fly to its new home. Both swarms showed the same set of events shortly before takeoff: the scouts became unanimous in their dancing and then the non-scouts began to move around excitedly. Both swarms also showed several striking changes in the images recorded with the infrared video camera. First, we found that during the last 15 minutes before the moment of takeoff, the percentage of the mantle bees with a hot (at least 95°F/35°C) thorax rose from about 5 percent to 100 percent. This change in the bees' thorax temperatures is shown in Figure 6.3. Second, we saw (in both swarms) that the bees took flight at the very moment that *every bee* in the swarm's mantle had a thorax temperature of at least 95°F/35°C. Moreover, we observed (after the mantle/outermost bees had taken flight) that the bees

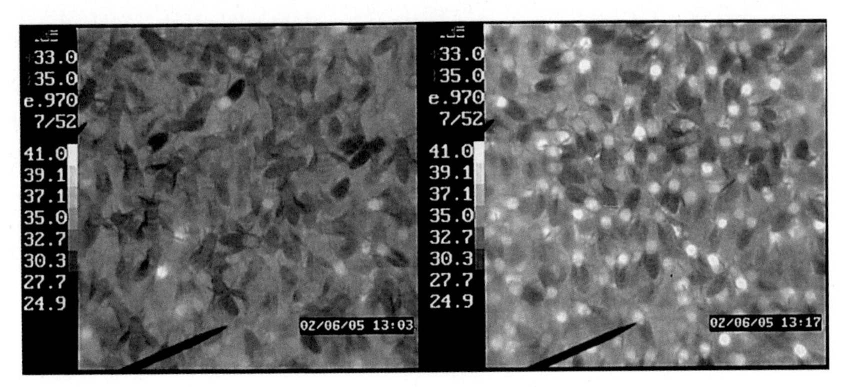

FIG. 6.3. Images showing the thorax temperatures of the bees on the surface of a swarm when viewed with an infrared video camera. *Left*: Image made 15 minutes before takeoff. *Right*: Image made 1 minute before takeoff. On the left side of each image is a scale that indicates the temperature (in °C) represented by white, black, and shades of gray. 24.9°C = 77°F, and 41.0°C = 106°F.

inside the swarm's cluster also had "white-hot" thoraces, that is, ones with temperatures at or above 95°F/35°C. This finding of high thorax temperatures in *all the bees in a swarm* explains something that is familiar to beekeepers: a clustered swarm needs only about 60 seconds to "disintegrate" (take flight) when it starts to move to its new home.

At this point you may be wondering, what stimulated the surface-layer (mantle) bees and everybody else in these swarms to warm themselves up? And how was it that each swarm's takeoff began just seconds after all of its bees had warmed their flight muscles to at least 95°F/35°C? In other words, what stimulated the swarm bees to prime themselves for flight, and what triggered them to take flight? Answer: the nest-site scouts, producing two special signals.

As mentioned above, back in the summer of 1970, I had heard high-pitched sounds coming from Doc's swarms shortly before takeoff, and I had likened them to the rising engine whine of a race car accelerating in a straight stretch of the racetrack. It seemed likely to me, therefore, that these high-pitched piping sounds were signals from the nest-site scouts to the other bees in a swarm, and that their message was "Ladies, warm your flight muscles!" But back in 1970, I never was able to identify which bees made these high-pitched piping sounds, because they seemed to come from inside the swarm cluster, thus from bees that were out of sight.

I discovered which bees in a swarm's cluster make the piping sounds in August 1999, thus 29 years after I first heard them. This discovery came by chance as an observation that I made at my cabin beside Ox Cove, up in Pembroke, Maine. I had set up a swarm (on my swarm board; see Figure 4.4) and had labeled with colorful paint dots the first dozen or so bees that performed waggle dances on this swarm. These bees were nest-site scouts. My aim was to watch closely the behavior of my color-coded scout bees, as part of a study of how the dissent among the scouts in a swarm gradually expires during the process of choosing the swarm's new home. On 2 August 1999, at 10:48 a.m., just five minutes before my swarm took off, my attention was drawn to scout bee Blue, who did something unexpected on the swarm's surface: she ran excitedly over other bees for a few seconds, then paused for about a second, grabbed a stationary bee and pressed her thorax against this bee, and then ran on, repeating the sequence of run-pause-press six times before she burrowed into the cluster and went out of sight (Fig. 6.4). I noticed that each time my scout bee Blue paused and grabbed another bee, she drew her wings tightly together over her abdomen and then she vibrated them slightly. Was scout bee Blue producing the piping sound? I heard this sound each time she had grabbed another bee, but with just my "naked" ears I could not be certain that the sound came from her. I needed a bee stethoscope.

That afternoon, I drove to Morgan's Garage in Pembroke and bought a 3-foot (1-meter) length of rubber vacuum hose about one quarter inch (ca. 6 mm) in diameter, a size that fit snugly in my ear. I figured that I could use this as a sound tube to localize the sources of sounds coming from my swarms. A few days later, when I watched a second swarm and used my rubber hose to listen in on another nest-site scout doing the run-pause-press behavior, I was thrilled to hear the exact same piping sound

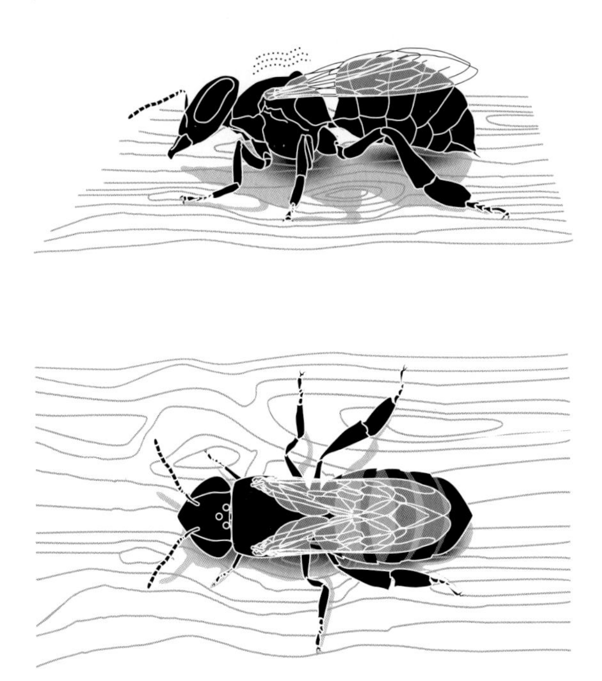

FIG. 6.4. A nest-site scout producing the piping signal. During a pause from running over bees in a swarm cluster, she presses her thorax to the substrate (show here as a wooden surface), pulls her wings together over her abdomen, and activates her flight muscles to generate a vibration in the substrate. In nature, the substrate will be another bee. A nest-site scout will produce piping signals on the swarm cluster when she senses that a quorum of scouts has built up at "her" site.

that I had heard 29 years before, while keeping watch over Doc's swarms atop Mount Pleasant.

I was fascinated by the sights and sounds of the piping bees in swarms. I was especially keen to describe their signal in detail and to test the hypothesis that nest-site scouts use this piping signal to stimulate all the other bees in a clustered swarm to warm their flight muscles in preparation for taking off to fly to their new home. I was able to do these things when Professor Jürgen Tautz, from the University of Würzburg, in Germany, joined me in Ithaca in August 2000 to investigate the piping signals produced by some of the worker bees in a swarm cluster. He came equipped with the miniature microphones and digital audio and video equipment we needed. We set up a swarm on my swarm board so that we could easily monitor everything that happened on the swarm's surface. Next, we installed two microphones and two temperature probes in the cluster of swarm bees. One microphone and one temperature probe were mounted in the swarm cluster's core; the other microphone and temperature probe were positioned on the swarm cluster's surface. There was also a third, handheld microphone for listening to the sounds produced by individual bees walking across the surface of the swarm. Directly in front of the swarm, we positioned a video camera that recorded both the sounds from the swarm's interior and the sounds from the individual bees that we tracked on its surface. With all the microphone and thermometer wires leading from our swarm, a video camera continuously recording its sights and sounds, and two biologists monitoring its temperatures, our swarm looked rather like a patient in an intensive care unit.

By now I had a precise search image for a piping bee—one dashing over the swarm's surface but pausing frequently to seize a motionless swarm-mate—so I was able to spot pipers at a glance when we started to hear their shrill sounds. From our video recordings, Jürgen and I quickly confirmed the discovery that I had made the summer before while watching a swarm in Maine: *piping bees are exceptionally excited nest-site scouts*. These bees demonstrated this fact to Jürgen and me by switching between worker piping and waggle dancing while they scrambled over the surface of the swarm, as shown in Figure 6.5. This mixing of signals became especially noticeable during the last half hour before takeoff, when the piping grew strongest.

At this point, Jürgen and I wanted to test the hypothesis that the function of worker piping is to stimulate all the bees in a swarm to prepare for takeoff. The main preparation they need to do is warming up their flight muscles, so they will be able to launch into flight. We tested this hypothesis by seeing whether or not worker piping occurs only in the last hour or so before takeoff, when all the bees in a swarm are making flight preparations. So, we measured simultaneously the level of piping in a swarm and the temperatures in its core and mantle (outer layer of bees)

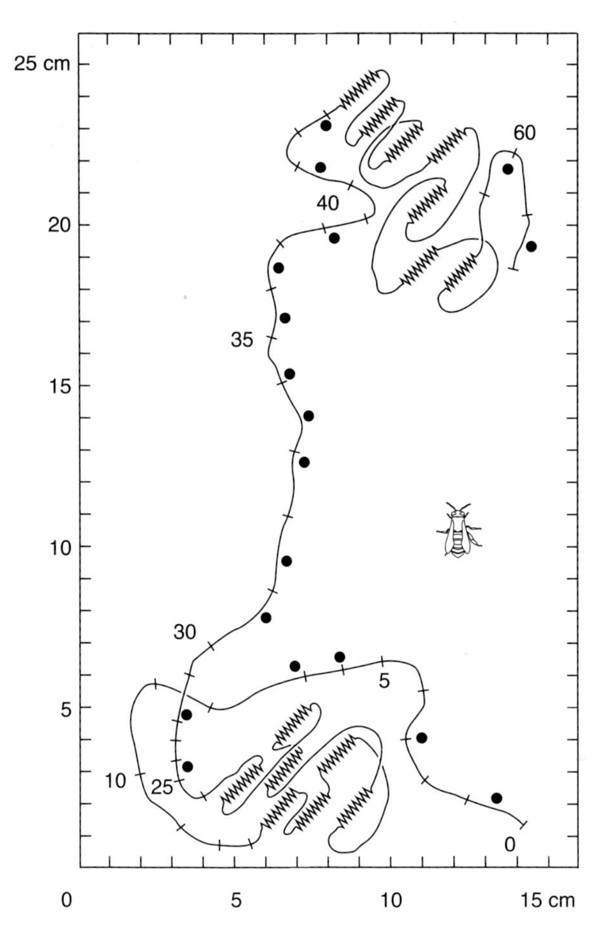

FIG. 6.5. Record of a scout bee switching between worker piping and waggle dancing as she ran over the surface of her swarm's cluster. Tick marks along her track denote 1-second intervals. Black dots mark worker piping and zigzags mark waggle dancing. The start of this 62-second-long recording of this bee's behavior began 2 minutes and 45 seconds before her swarm launched into flight.

over many hours prior to takeoff. Figure 6.6 shows an example of the patterns in piping and warming that we found. Three hours before takeoff, when the ambient temperature was 73°F (23°C) and the swarm's core and mantle temperatures were 93° and 87°F (34° and 31°C), we heard no piping. Then, about two hours before takeoff, we began to hear worker piping, but only intermittently. Finally, during the half hour before takeoff, the sound of the piping workers was loud, for by then numerous bees

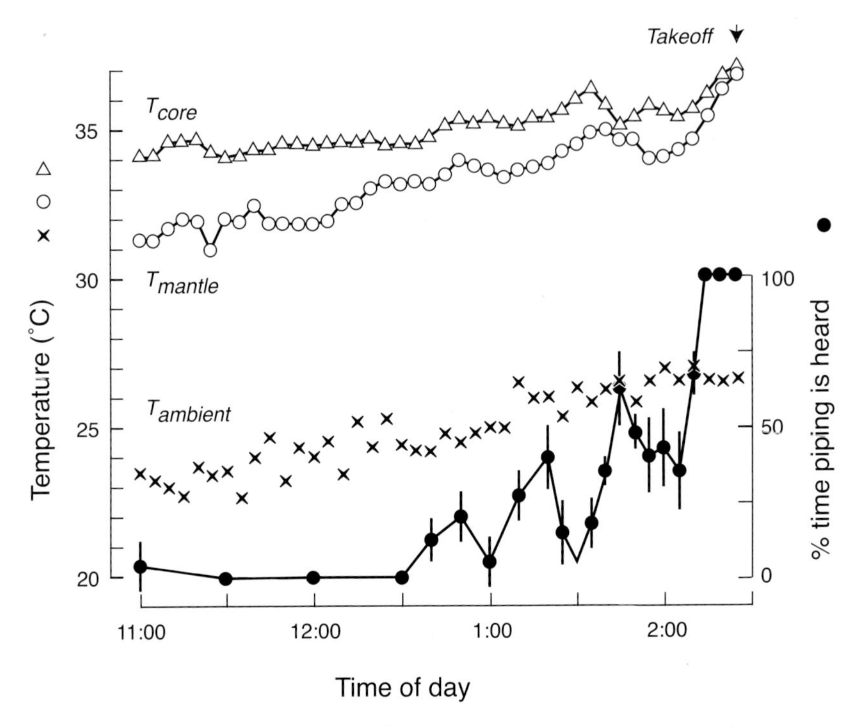

FIG. 6.6. Pattern of worker piping (filled circles), swarm temperatures (open circles and triangles), and ambient temperatures (x's) during the three-hour period preceding a swarm's takeoff. This figure shows that the workers' production of piping signals and the swarm's temperature rose together over the hour before takeoff. Only the nest-site scouts from the winning site produce piping signals. As explained in Chapter 5, they do so after sensing that a quorum of fellow scouts has built up at their site.

were piping simultaneously. At the same time, the temperature in the mantle was rising rapidly, and just when the temperature throughout the swarm reached 99°F (37°C), the bees launched into flight! The fact that worker piping coincides perfectly with swarm warming—both phenomena rise to a climax at takeoff—shows that the worker piping signal functions to stimulate all the bees in a swarm to prepare for takeoff.

If you are a beekeeper, then it is *very* useful to know about the piping signal. It is one of two things—hearing piping bees and seeing buzz-running bees (discussed in the next chapter)—that will tell you that a swarm that is hanging from a tree branch, a stop sign, a picnic table, or whatever, is preparing to launch into flight. So, put one ear within a few inches of the swarm and listen. If you hear a high-pitched chorus of nest-site scouts producing their shrill piping signals, then you'll know that this swarm's bees are still getting ready for their liftoff *en masse*. It will happen soon, but not right away. But if you hear loud buzzing *and* you see bees running wildly across the swarm, then you'll know that the liftoff is imminent, as we will see in the next chapter. But there are things you can do to prevent the swarm from flying off. One is to shake it immediately into an empty hive. Another is to spray it lightly with water. This will cool the bees and buy you enough time to don a veil and calmly and safely present the bees with a fine home . . . your empty hive!

CHAPTER 7

Boisterous Buzz-Runners

The first substantial book on beekeeping in the English language is *The Feminine Monarchie*. It was "written out of experience" by the Reverend Charles Butler and was published in 1609. Butler was a keen and observant beekeeper. He kept some 40 colonies in straw skeps around his parsonage in the town of Basingstoke, which sits about 45 miles (70 kilometers) west of London. In writing his book, Butler drew mainly on what he had seen firsthand, and by doing so he corrected various misunderstandings about the lives of honey bees. For instance, he confirmed that the "king bee," described by countless writers as the driving force in a honey bee colony, is in fact a queen bee. He also deduced that honey bees produce (rather than collect) the wax for their combs, that drones are males, and that workers are females. My favorite passage in *The Feminine Monarchie* is Butler's description of how the appearance of a bivouacked swarm changes just before it launches into flight from its resting site: "And so doth this soft shivering passe as a watch-worde from one to an other, untill it come to the inmost Bees: whereby is caused a great hollownesse in the Cone. When you see them do thus, then may you bid them farewell: for presentlie they begin to unknit, and to be gone."

Honey bees present us with many wonderful sights, but I feel that what Butler describes here—how the bees in a swarm cluster "begin to unknit" to form a cloud of thousands of swirling bees—is *the most exciting sight* that a beekeeper can behold (Fig. 7.1). Anyone who watches this process

FIG. 7.1. The author in 2007 watching a swarm launch into flight from the vertical board that is used as a swarm mount. Two feeder bottles on the mount provide sugar syrup to keep the swarm bees well fed.

will marvel at how a bunch of bees can take off so synchronously and at just the right time. From start to finish, the process of some 10,000 bees launching into flight requires only about 60 seconds. And it does indeed start when the scout bees in a swarm produce a special signal that means "Let's go." These days, we call this "watch-worde" the "buzz-run."

Martin Lindauer, the German biologist whom we "met" in Chapter 3, was the first to describe the bees' takeoff signal in a scientific paper. He called it the *Schwirrlauf,* which translates nicely to English as the "buzz-run." Both names are apt, for when a worker bee produces this behavioral signal, she scrambles across the swarm cluster, moving in a zigzag pattern while noisily buzzing her outspread wings, sometimes running over the backs of the immobile bees, and other times bulldozing between them (Fig. 7.2). Lindauer also reported that bees performing buzz-runs become prominent on a swarm's cluster in the final minutes before a swarm starts its takeoff, and he suggested that by barging into and boring through the cluster the buzz-runners push their swarm mates apart and trigger their explosive launching. His interpretation of this boisterous behavior made a lot of sense to me, but I needed to see if it really fit the facts. Thus it was that in May 2005, exactly 50 years after the

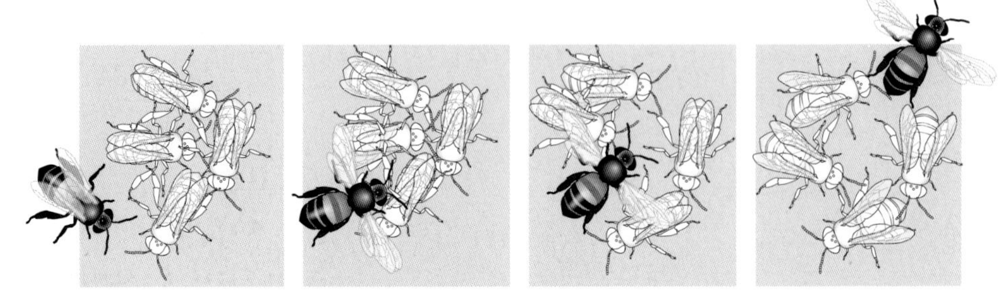

FIG. 7.2. Worker bee performing the buzz-run through a tangle of lethargic bees. Panel 1: The buzz-runner runs towards the quiet bees with her wings folded over her abdomen. Panel 2: One second later, the buzz-runner spreads her wings and buzzes them as she makes contact with the cluster. Panel 3: One second after making contact, the buzz-runner pushes through the cluster while still buzzing her wings. Panel 4: She breaks contact with the bees but continues buzzing her wings as she runs on. Based on 4 frames of a video recording.

publication of Lindauer's 1955 paper "House hunting by honey bee swarms," I began to take a close look at the form and function of the buzz-runners' behavior.

By May 2005, I had already studied the workers' piping signal, so I knew that worker piping functions as the flight-preparation signal that tells the bees in a swarm to warm up their flight muscles. Now I wanted to learn more about the buzz-run signal. I had recently acquired a new, state-of-the-art digital video camera and playback deck, thanks to support from the National Science Foundation. So, nearly 400 years after the publication of *The Feminine Monarchie*, the time had come for a close look at how "doth this soft shivering passe as a watch-worde from one to an other." I was keen to understand the interplay between worker piping and buzz-running, since both signals are produced when the nest-site scouts have finished choosing their swarm's future homesite. There were many questions. What is the precise form of the buzz-run behavior? What causes a bee to become a buzz-runner? What effects does a buzz-runner have on the bees she contacts? And which bees in a swarm produce the buzz-run signal?

A search of the scientific literature revealed just one study, besides Lindauer's 1955 work, on this topic. In 1967, Harald Esch, a student of

Karl von Frisch, published a study of the sounds produced by a buzz-runner. He found that when one is running around on a swarm, she produces short pulses of 180–250 Hz wing vibrations, and then shifts to delivering longer bursts of 500 Hz buzzing when she pauses to "blast" another bee with her message, "Get going!" There are no soft-spoken buzz-runners.

I was assisted in the study of these rabble-rousing buzz-runners by Clare Rittschof, an undergraduate student at Cornell who was keen to gain experience in studying honey bee behavior. Clare is now a professor at the University of Kentucky, where she studies how the environment in which a worker honey bee develops can influence her future behavior. For example, developing in a colony that is fiercely defending itself against robber bees can heighten, later in life, a worker bee's vigilance for bees trying to sneak into her colony's home.

Clare and I started our work by watching for buzz-runners on swarms. Our first goal was to get a clear picture of when they produce their attention-getting signals. To do so, we mounted a swarm of bees on one side of a vertical wooden board (like what is shown in Fig. 7.1) and then we video recorded the activities of all the bees on the swarm's surface that were within a 4×6-inch (10×15-centimeter) area on the "front side" of the swarm. We started our video recording when we heard piping signals coming from the swarm and we ended it when all the bees took off to fly to their new home. While we conducted the video recording, we watched for bees running over the surface of the swarm, and whenever we saw one we followed her for about 10 seconds with a small microphone, to eavesdrop on any sounds she might be producing. The microphone was connected to the video camera, so whatever sounds we heard while following a running bee were recorded for use in our analysis of how this worker bee behaved. We expected that some of the fast-moving bees would be pipers, others would be buzz-runners, and some might be both. We needed to know what signal(s) each bee produced. The buzz-runner bees were easily distinguished acoustically from the piper bees by the loud bursts of buzzing sound whenever a buzz-runner spread her wings for a second or so. Clare then spent several weeks playing back our video recordings in slow motion, scanning them for bees running over the swarm's surface and watching most closely those bees that she identified as buzz-runners based on their signature buzzing sound.

Clare's careful studies of our swarm surveillance videos yielded two key findings. First, she saw that more and more bees began running excitedly (and erratically) across our swarms during the final hour preceding takeoff, so that just before a swarm launched into flight it teemed with bees dashing over and through the cluster. Second, and more surprisingly, she saw that *all of the running bees produced audible signals*: pipings or buzzes, or both. Initially, most of the running bees produced just piping signals. But as time passed, more and more of the bees that were scrambling over the swarm's surface performed buzz-runs, and sometimes even rammed themselves into other bees while revving their wings. Clare's video analysis was especially revealing about the last five minutes before takeoff; she found that nearly 70 percent of the fast-moving bees produced pipings *and* buzz-runs, and 24 percent produced buzz-runs only (Fig. 7.3). This showed us that most—probably all—of the buzz-runners are the same bees as the pipers. This was an important finding, because we knew already that the pipers are nest-site scouts. So, now we knew that it is the nest-site scouts in a swarm that produce both the piping signals to *prime* the swarm for takeoff and the buzz-run signals to *trigger* its takeoff.

There is a feature of the behavior of buzz-runners that is important to note: sometimes a buzz-runner will launch into flight, fly around the swarm cluster for a few seconds, and then land back on the cluster and resume her buzz-running. This phenomenon of buzz-runners taking flight is important because it points to the evolutionary origins of this signaling behavior. Almost certainly, the buzz-run signal is a ritualized form of a bee's takeoff behavior, which consists of a bee spreading her wings, starting to buzz them, pushing clear of other bees, and then taking to the air.

"Ritualization" is the term that biologists use to refer to the process whereby some *incidental action* of an animal becomes modified over evo-

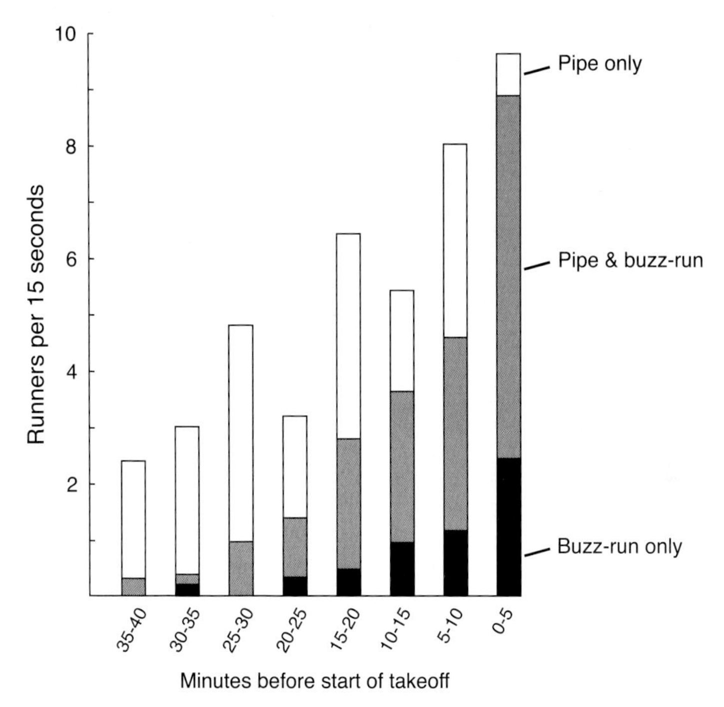

FIG. 7.3. The record of the increase in number of bees running across a fixed portion of a swarm's surface (within 15 seconds) during the 40-minute period preceding a swarm's takeoff to fly to its new homesite. This figure also shows the changing pattern of signal production by the running bees. Initially, most of these bees produced just piping (the "Time to warm up!" signal), but in the last 15 minutes, most produced buzz-runs (the "Time to go!" signal), too.

lutionary time into an *intentional signal*. Often, the incidental action is a by-product of an activity performed in one particular situation. If so, then the incidental action is a reliable indicator of this situation. The honey bees' buzz-run signal illustrates this idea nicely: when a bee is about to take flight, she inevitably buzzes her wings, therefore wing buzzing by a bee is a reliable indicator to others that she is about to take flight. The next step in the evolution of a signal is for the receivers of it to use the information it provides to improve their decision making. If the receivers' improved decision making benefits the senders, then the senders will benefit by making the signal more conspicuous and thus more easily detected by the receivers.

In the early stages of the evolution of the buzz-run signal, the quiescent bees in a swarm's cluster probably benefited from sensing the wing buzzing of other bees, to know when they (the quiescent bees) should get ready to launch into flight. We do essentially the same thing when we are stuck in a traffic jam that stretches so far out of sight that everyone has turned off his or her car's engine. When, eventually, we hear the engines in cars ahead of us getting turned on, we know that we should turn on our car's engine, too. In the case of the bees, better decision making by the quiescent bees about when to warm up for takeoff probably produced more synchronized takeoffs, which also benefited the active bees. So natural selection favored exaggerations of the wing buzzing by the active bees to make it more conspicuous to the quiescent bees. Given the present-day form of the buzz-run, it looks like these exaggerations include starting the wing buzzing long before the moment a buzz-runner takes flight, plus adding the actions of running over and ramming into other bees.

Now for a "thought question" regarding the buzz-run: Why do honey bees have this signal? In other words, why do the non-scouts in a swarm wait to receive the buzz-run signal from the scouts before they launch into flight? I think that the non-scouts need to wait for the buzz-run signal from the scout bees because it is only the scout bees that run across a swarm cluster, so it is only the scout bees that are able to know when all the bees in the swarm are warmed up and ready for takeoff. The buzz-run signal enables the scouts to share this critical piece of information with their swarm-mates who do not run around. As we have seen in Chapter 6, for all the bees in a swarm to launch into flight at the same time, they must wait to do so until every bee has her thorax warmed to around 97°F (35°C). So, how can *all* the bees in a swarm know when they've *all* become hot enough? I think that the only good way is to have some bees (the scouts) scurry over the swarm cluster, sense the temperatures of their swarm-mates along the way, and then produce a "time to go" signal (the buzz-run) when their canvassing tells them that a sufficiently high body temperature has been reached by all the bees.

The evidence shows that this is how it works on swarms. We now know that once the scout bees at the winning site have sensed a quorum of scouts there, they return to the swarm and begin to scramble over and into the cluster, pausing every few seconds to produce the piping signal. Each time a scout presses against another bee to pipe to her, she can sense the other bee's body temperature. Eventually, once the swarm cluster is fully warmed up, she will press against bee after bee that is hot enough to take flight, and this tells her that it is time to start producing the buzz-run signal. This is why we see that the production of the buzz-run signal crescendos in the final minutes before takeoff, right when all the bees have warmed their flight muscles to the high temperature needed for flight. Reminder: as noted at the end of Chapter 6, if you see buzz-runners scrambling over the surface of a swarm, and you want to prevent the swarm from taking off so you can shake it into a hive, then just spray the swarm lightly with water (ideally, by misting it with a spray bottle) to cool the bees. This will "buy" you the time needed to hive the swarm.

In the following chapters of this book, we will see that for all the rest of a colony's activities—for example, its comb construction, food collection, and nest thermoregulation—the control system is a highly distributed process involving thousands of bees. But in these chapters, we have seen something fundamentally different. It is a *small minority* of a swarm's members, the nest-site scouts, who survey the group to collect information about its temperature, and then produce the signal that triggers everyone to take flight at just the right time. Sometimes, I struggle to believe that worker honey bees—"mere" insects—possess the behavioral sophistication that is needed to perform all the things that nest-site scouts do: find and inspect potential homesites, advertise them appropriately, conduct a sophisticated "evidence race" to identify the best one, organize the swarm's takeoff, and finally, direct everyone during their move to the new homesite. In the next chapter, we will examine the last item in this set of the amazing abilities of nest-site scouts: steering the swarm to its future dwelling place.

CHAPTER 8

Flight Control in Swarms

Anyone who watches a honey bee swarm launch into flight and move off to its new home is presented with a mind-boggling puzzle: how does a school-bus-size cloud of some 10,000 bees manage to fly directly to its new dwelling place? Its flight route might extend for miles, so the swarm may pass over fields and woods, hilltops and valleys, and maybe even a wetland or river. Also, its destination is extremely specific, such as a small knothole in a big tree that is deep in a forest. There can be no doubt that a swarm has precise control of the direction and the distance of its flight. Furthermore, as we shall see, a swarm also carefully controls its flight speed. When a swarm nears its destination, it slows down gradually and then stops just outside the "front door" of its new home.

In the summer of 1972, while working at the Dyce Lab, I had the good fortune of being able to watch several swarms choose a future home site, launch into flight, and then set out on their journeys to their future dwelling places. This was a summer when Doc was joined by Professor Alphonse (Al) Avitabile from the University of Connecticut Waterbury. Both Al and Doc were fascinated by the behavior of honey bee swarms, and that summer they teamed up to investigate how the workers in a swarm sense the presence (or absence) of their queen when a swarm flies to its new home. My work assignment from Doc was to be their helper. This meant that I helped watch the swarms prepare for their flights, and I made notes on what happened once each swarm had taken to the air.

The background to this study was work done by Colin G. Butler, of the Rothamsted Experimental Station in England. He had discovered that when worker bees are inside their hive they monitor the presence of their queen by smelling the "queen substance" pheromone that she produces in special glands in her head (to be discussed in Chapter 10). The main component of this pheromone is a 10-carbon fatty acid. Its full name is (E)-9-oxo-2-decenoic acid; its abbreviated name is 9-ODA. Al and Doc wanted to see if the workers in an airborne swarm also sense the presence of their queen by smelling her signature pheromone (9-ODA). It was known already that if the workers sense her absence—usually because she has dropped out to rest—then they will stop flying forward, will search for her, and soon will assemble around her wherever she has landed. I suspect that this is why honey bee swarms sometimes settle in strange places where they are not welcome, such as traffic lights, store signs, mailboxes, and parked aircraft.

To test the hypothesis that the scent of 9-ODA is the indicator of the queen's presence in an airborne swarm, Al and Doc conducted an experiment that was both simple and revealing, hence it was elegant. They set up seven swarms, one at a time, on a wooden cross that stood in the hayfield just south of the Dyce Lab. Each swarm's queen was confined in a small cage that was lashed to the cross, as described in Chapter 6. Then, when each swarm had finished choosing its future homesite and was about to take off to move there—that is, when we began to hear the then mysterious piping sound coming from the swarm cluster—Al put a drop of a saturated solution of 9-ODA in distilled water on five of the swarm's worker bees. Then what happened? Well, six of the seven swarms flew away over the woods to the north and the east of Dyce Lab, so we soon lost sight of them. The seventh swarm, however, flew off to the southwest across recently mowed hayfields. We managed to run beneath this swarm and accompany it all the way to its new homesite: a cavity in a white pine about 1,000 feet (300 meters) away.

The revealing part of this experiment occurred next, when we flew another seven swarms that were treated in exactly the same way as the first seven, except there was no 9-ODA in the water that Al applied to five worker bees in each swarm just before it launched into flight. We saw that all of these swarms took off and started to fly away, but that after moving off about 150 feet (50 meters), the bees in each swarm dispersed and then drifted back to their caged queen, where they resettled around her. It was clear that the workers in these swarms had sensed that their queen was missing. These results showed us that it is the presence of the scent of 9-ODA that informs the worker bees in an airborne swarm that their queen is "on board." These results also taught me a memorable lesson about scientific work: sometimes, all you need to answer your question is one incisive experiment.

Several years later, in 1979, I returned to the study of flight control in honey bee swarms. This time, Doc, Kirk Visscher (who was starting graduate studies in the entomology department at Cornell), and I conducted a study that gave us a clear picture of the flight of a swarm, from start to finish. We did so by taking an 11,000-bee swarm out to Appledore Island in Maine. I knew from my previous work on the island that this nearly treeless location could give us good control of our swarm's flight path to its new home. We had two goals: (1) measure the flight speed of our swarm at each stage of its journey, and (2) measure the percentage of our swarm's bees that had already visited the new homesite (and so knew its location) when the swarm took off to fly to it.

Our first task upon arriving on Appledore was to find a flight route that we could run along to watch our swarm throughout its move. Finding this route was tricky. Thickets of poison ivy (*Toxicodendron radicans*) cover much of Appledore Island, and steep-sided granite ledges stretch across it. We settled on a 1,902-foot (580-meter) flight route that ran near some of the island's foot paths and dirt roads; we hoped this would enable us to run with the swarm throughout most of its flight. This route stretched from an open spot near the dock on the island's west side to a grassy spot above the cliffs on its east side (Fig. 8.1). We tied orange flagging to shrubs spaced 98 feet (30 meters) apart along the first 1,280 feet (390 meters) of the flight route, and we crossed our fingers that the swarm

FIG. 8.1. *Top*: Appledore Island, Maine, seen looking south from an airplane. The dashed yellow line indicates the 1,902-foot (580-meter) line along which we monitored the flight of a swarm moving from its cluster site to a bait hive. *Bottom*: Our swarm, clustered on a wooden cross with a bottle of sugar syrup to keep its worker bees stuffed with food, hence well fueled for the flight to its new home.

would travel straight down this route. (Scary-looking thickets of poison ivy kept us from extending our string of orange flags beyond 1,280 feet.) Then we positioned our bait hive at the eastern end of the planned flight line. Finally, we set up our swarm on a wooden cross.

This swarm consisted of approximately 11,000 worker bees, one queen, and probably some drones. We brought these bees to the island in a "package cage," i.e., a cage made of wood and hardware cloth and used to send a honey bee colony through the mail. We induced our swarm to cluster on our cross by lashing to it the little cage that held the swarm's queen, and then shaking the workers and drones out of the cage and onto the ground at the foot of the cross. The bees flew up and crawled up and formed a cluster around their queen. Once they had done so, we released Her Majesty from her cage. At this point, we knew that whatever would happen next would be up to the bees. We crossed our fingers that the nest-site scouts in our swarm would find and choose our bait hive (not Rodney Sullivan's chimney!) and then would steer the swarm to fly straight down our flight route. If the swarm did so, then we would be able to record (1) when it took off, (2) when the center of its "cloud" passed over each 30-meter waypoint, and (3) when it arrived at our hive. This way, we would be able to calculate its speed during each stage of its flight.

We were lucky. Everything proceeded smoothly. The scout bees in the swarm chose our bait hive for their future home, and they piloted the swarm straight down the beeline flight route to it. We, in turn, managed to scurry along beside the cloud of swirling bees throughout its 6.4-minute flight except for the last segment, when it flew over the no-man's land of a poison ivy jungle. We learned that our swarm reached its top speed—5 miles per hour (8 kilometers per hour)—after traveling about 500 feet (150 meters), and that it began to slow down when it was about 300 feet (90 meters) from our hive.

We also learned what percentage of the bees in the swarm visited our bait hive before the swarm flew to it. To measure this percentage, I put a dot of yellow paint on each scout bee that advertised our hive, that is, each bee that performed a waggle dance for it on the swarm cluster. Meanwhile, Kirk and Doc recorded at the hive what percentage of the bees scouting it bore a dot of yellow paint. Knowing that I had painted 143 bees, and that 29 percent of the scout bees sighted at the hive bore a dot of yellow paint, we calculated that approximately 495 bees ($143 = 0.29 \times 495$) from our swarm had visited the hive before the swarm made its takeoff. This showed us that fewer than 5 percent of the bees in our 11,000-bee swarm knew its destination when the swarm took off to fly to its new residence.

We remained in the dark, however, about two key things: (1) how the scout bees steered the airborne swarm to its new dwelling place, and (2) how these bees "applied the brakes" when the swarm approached its destination. In other words, we knew a good deal about the *pattern* of our swarm's flight, but we knew rather little about the *processes* whereby the scout bees had controlled its flight.

Neither I nor anyone else pursued the mysteries of flight control in honey bee swarms for the next 25 years. This hiatus ended in 2004, when Professor Madeleine Beekman, from the University of Sydney, in Australia, joined me and two Cornell undergraduate students, Adrian Reich and Robert Fathke, for a summer of honey bee research. We decided to describe further the flight patterns of swarms and to perform experiments on their mechanisms of flight guidance. We also decided to try to do these studies at my bee lab at Cornell, rather than out on Appledore Island.

My bee laboratory is part of the Liddell Field Station (LFS), a facility that supports laboratory and field studies by biologists from various departments at Cornell University, but especially the Department of Neurobiology and Behavior. The LFS sits about two miles from the center of the Cornell University campus, and it is surrounded by many hundreds of acres of forests and fields, all owned by the university. The forests provide natural areas for hikers to enjoy, and the fields support various research programs, including mine. In 2004, Madeleine and I made good use of the 104-acre (42-hectare) hayfield that lies just to the west of the

FIG. 8.2. The lone tree in the large field just west of my laboratory at Cornell University.

LFS. In the center of this field stands just one tree, a large white ash (*Fraxinus americana*), shown in Figure 8.2. This tree was the perfect place for mounting a bait hive for our studies of swarm flight guidance.

There were, of course, other good homesites for honey bees in the woods nearby, but by 2004 I had learned that if I steadily watched the dancing bees on the surface of a swarm and removed *every* bee that I saw performing a waggle dance for a site other than my bait hive, then I could keep the attention of the swarm's scout bees focused on my bait hive. Using this technique, Madeleine, Robert, Adrian, and I performed experiments in which swarms flew along an 886-foot (270-meter) flight route that ran west from the eastern edge of the field shown in Figure 8.2 out to the ash tree. We planted stakes along this flight route at 98-foot (30-meter) intervals to make measurements of the flight speeds of our swarms. We also mowed clear a 66 × 66-foot (20 × 20-meter) "launch pad" for our swarms. Within it, we created a checkerboard array of stakes spaced 13.2 feet (4 meters) apart, and beside the swarm cross we erected a tall "ruler" (built of PVC pipe) that stood straight up nearly

20 feet (6 meters). These size references helped us judge accurately the width, length, and height of a swarm's "cloud" shortly after liftoff.

We began our study by describing the flights of three swarms. Each one contained approximately 12,000 bees, which is the average size of the 235 swarms that Doc and his students collected in Ithaca in the mid-1970s. We mounted our three swarms, one by one, on the swarm board that I like to use when studying swarms (Fig. 4.4). When each one finished choosing our nest box for its new homesite, and then launched into flight to move to it, we saw that its cloud of airborne bees measured about 33 feet (10 meters) long, 26 feet (8 meters) wide, and 10 feet (3 meters) tall. In each case, the bees at the bottom of the cloud flew about 6 feet (2 meters) above the top of the grass. Handy headroom for us! Knowing these dimensions, we calculated the bee density in each swarm's cloud. For all three swarms, it was approximately 1.4 bees per cubic foot (50 bees per cubic meter).

We found that the flight patterns of our three swarms closely matched what Doc, Kirk and I had observed 25 years before with our swarm on Appledore Island. At first, each swarm cloud moved slowly toward its destination. Then it accelerated, and reached a top speed of around 4 miles per hour (7 kilometers per hour) after it had flown about 500 feet (150 meters). When it came within about 300 feet (90 meters) of its destination, it began to put on the brakes (Fig. 8.3). Finally, as soon as the swarm bees reached the nest box, they began to land on it and scramble inside. Within about 10 minutes, almost every bee was out of sight, safe inside the colony's cozy new residence.

Each time I watched the bees in our swarms arrive at and march into their new home, my thoughts jumped back 41 years to my memories of watching the bees in a swarm doing the same thing, but on the lowest limb of the massive black walnut tree near my boyhood home. And I would smile to myself, still admiring the orderliness of it all. I hope that someday you too will have an opportunity to watch a swarm of bees flying over a field, and maybe even arriving at and moving into its new homesite. These sights are unforgettable.

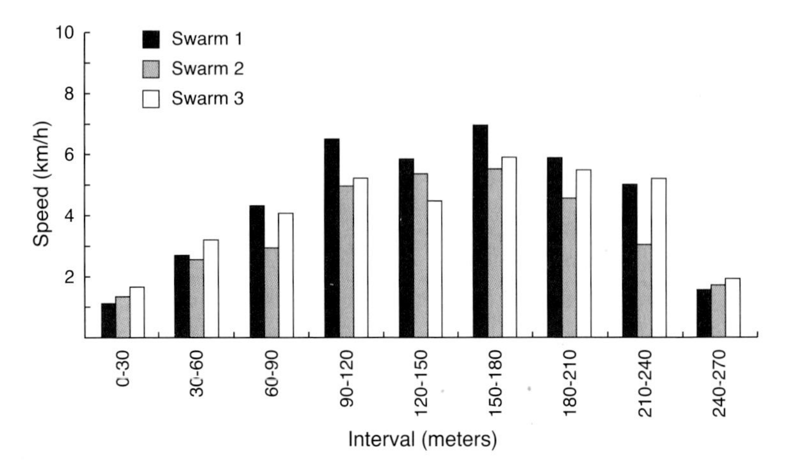

FIG. 8.3. Flight speeds of three swarms during every 99-foot (30-meter) stage of their flights from the swarm board to the nest box. Top speeds were 3–4 miles per hour (5–7 kilometers per hour). When swarms make longer flights, they can reach speeds greater than 7 miles per hour (12 kilometers per hour). This figure shows how an airborne swarm achieves a smooth acceleration from the start of its flight and a gradual slowing toward the end of the flight.

Madeleine and I next set ourselves the goal of testing the idea that the scout bees in a swarm guide the other bees using the attraction pheromone that they produce in their scent organs. Doc had stated in his 1975 paper with Al Avitabile and Rolf Boch (a fellow bee researcher based at the Ottawa Research Station in Ontario, Canada) that "the airborne swarm together with its queen, is led by worker bees releasing Nassanoff [*sic*] pheromone." It seemed to me that this statement was based on razor-thin evidence: "5 natural swarms whose queens were free flying . . . were 'lead' easily by the artificial Nassanoff scent." But, I wondered, where were they led? In what setting? And how far? It was clear that a more rigorous investigation was needed.

Our plan for examining this matter more carefully was to prepare swarms in which *every* worker had her scent organ sealed shut and then see if these swarms would perform well-oriented, full-speed flights down our flyway in the hayfield near my laboratory. The scent organ of a worker honey bee lies on the dorsal surface of her abdomen, just in front of her

FIG. 8.4. Worker bee exposing her scent organ (Nasonov gland) by tipping down the last segment of her abdomen. She is also using her wings as impellers, driving airflow to disperse the volatile pheromone from the Nasonov gland.

last abdominal segment. Usually, this surface is concealed by the last two plates ("tergites") covering the top of the abdomen, but if a worker bends the last segment of her abdomen downward, she exposes the membrane of the scent organ (Fig. 8.4). This releases the lemony aroma of the bee's assembly pheromone, whose message is "Come hither!" Using a tiny paintbrush, one can put paint over the joint between the last two tergites. Then, when the paint dries, they will be glued together so the bee will not be able to expose her scent organ. Aided by Robert Fathke and Adrian Reich, Madeleine and I performed this operation over and over until we had the tips of the abdomens of 4,000 bees properly painted. This was enough bees for a small "treatment" swarm. To control for any effects of chilling, painting, and handling the worker bees, we also prepared a 4,000-bee "control" swarm in which we did everything the same except that we applied the dot of paint to each bee's thorax instead of her abdomen.

Late in the summer of 2004, our team prepared, and then flew, six swarms down our special flyway at the Liddell Field Station: three treatment swarms and three control swarms. We found no differences in flight performance between the swarms in the two treatment groups. Their clouds of flying bees did not differ in size. Also, all six swarms flew directly and quickly to the nest box. The two types of swarm did differ in behavior, however, *when they reached the nest box*. We found that the treatment swarms took longer than the control swarms to move into the box: 20 minutes vs. 9 minutes, on average. And we saw why: the treatment swarms were hampered at the final, move-in stage because their scout bees could not guide the non-scouts to the entrance to their new home by marking it with Nasonov gland pheromones. We concluded that the scouts in a swarm do not use these pheromones to give their non-scout sisters flight-guidance information, but that they do use these pheromones to help their non-scout sisters find the entrance to their new home.

The last stage in these studies of flight control in swarms came in the summer of 2006, when I worked with Kevin Passino, a whiz professor in the Department of Electrical and Computer Engineering at Ohio State University, and Kevin Schultz, an ace graduate student working with Professor Passino. I met Kevin Passino when I made a trip to Ohio State University in 2002 to give a lecture about the flights of honey bee swarms. After my presentation, we talked, and I learned that he designed control systems and was intrigued by the mystery of flight control in honey bee swarms. In Kevin's engineering lingo, honey bees have evolved a "cooperative control strategy for groups of autonomous vehicles." We were keen to work together. So he, Kirk Visscher (then a professor in the Department of Entomology at the University of California, Riverside), and I went out to Appledore Island in the summer of 2006 to investigate further how the nest-site scouts control the flight direction of a swarm. Kevin figured that what we should do was video record (from below) a swarm flying to its new home. He explained that we could use point-tracking algorithms invented by engineers working on computer vision to track individual bees as the swarm passed over the video camera. These algorithms would determine, for each bee, her height in the swarm cloud, her flight speed, and her flight direction. Kevin also explained that the analysis would work best if the bees were flying in perfectly still air.

Appledore Island is a windy place. Indeed, it is so windy that there is now on the island an 80-foot (24-meter) tall wind turbine that supplies most of the electricity needed by the Shoals Marine Laboratory. Luckily, the air was nearly dead calm on two days during our stay there in 2006: June 29 and July 2. So, twice we were able to get a swarm to fly directly over Kevin's high-speed video cameras positioned at the 50-foot (15-meter) mark and the 200-foot (60-meter) mark down the flight line to a nest box that sat 650 feet (200 meters) from its takeoff site. Our video recordings were then processed using software created by Kevin Schultz. His computer program tracked, frame by frame, each ellipsoidal blob (i.e., each honey bee) in the recorded images. This tracking revealed the trajectory of every bee that appeared in the visual field of the video camera when the swarm flew over it. Also, the size of each ellipsoidal blob told us the height of the bee above the camera, so we could distinguish the movements of bees in the top and the bottom of the swarm cloud.

I struggle to express in words what it was like to go from watching thousands of swarm bees swirling overhead, to seeing graphs that revealed clear patterns in the commotions of their flights (Fig. 8.5). The words "wonderful" and "amazing" describe only a bit of my joy. For me, the most delightful revelations of this analysis of the video recordings are the proofs that the fastest-flying bees were indeed streaking in the direction of the chosen homesite (with a "bee flight angle" of 0°), and that the slowest-flying ones were heading in the opposite direction (180°). Also, comparing the upper and lower plots in Figure 8.5 shows that the speediest bees were mainly in the upper half of the swarm cloud. These results, confirmed subsequently by an independent study conducted in Germany, support the hunch that the nest-site scouts in a swarm steer

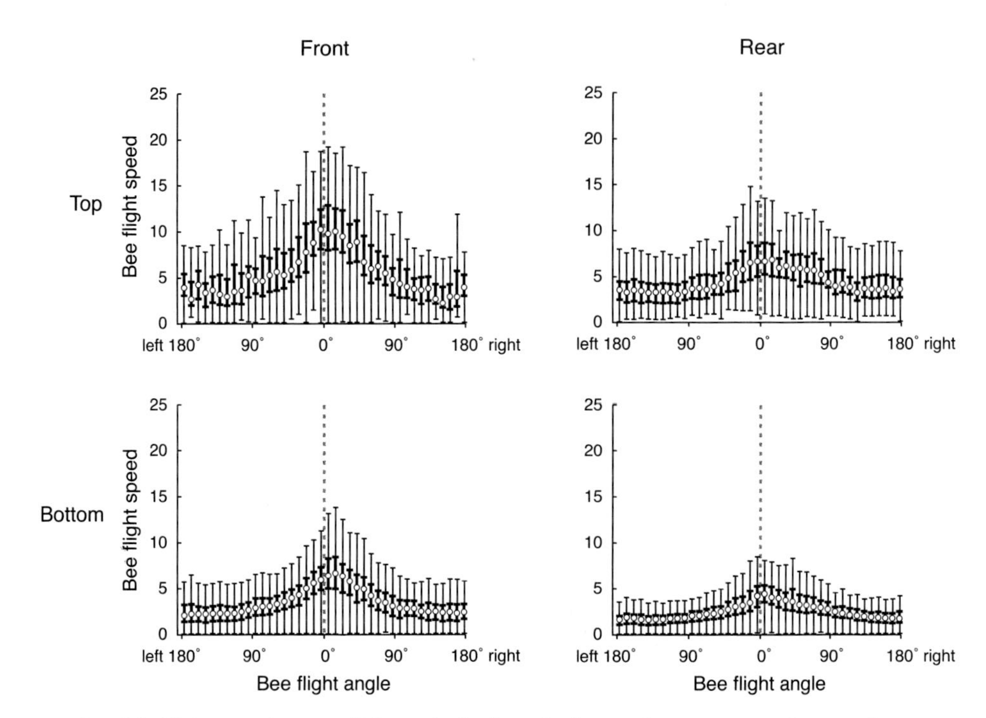

FIG. 8.5. Flight speed versus flight angle for bees in the top front, top rear, bottom front, and bottom rear of the cloud of an airborne swarm when it had flown ca. 50 feet (15 meters) from its bivouac site. A bee with a flight angle of 0° was flying straight toward the new homesite. The units of "bee flight speed" are bee lengths per video frame. This figure shows that the fastest-flying bees were in the top of the swarm cloud and were zooming straight toward the swarm's destination, so it is likely that many of these bees were nest-site scouts. This figure also shows that the fastest-flying bees accelerated as they flew through the swarm cloud from rear to front; their flight speeds were higher in the swarm's front than in its rear.

everyone else by streaking through the top of the swarm cloud, thereby pointing the way to their new home. But how the bees in a swarm cloud manage to "apply the brakes" when they get about 300 feet (90 meters) from their destination (see Fig. 8.3) remains a mystery. I have no doubt that here, too, the wondrous nest-site scouts are in control.

CHAPTER 9

Astonishing Behavioral Versatility

For many years, I thought that the bees in a colony that is preparing to swarm begin the search for their swarm's new homesite only after the swarm bees have left their original home and have assembled in a cluster nearby. I thought this because I figured that it is only after a swarm has separated out from its old colony that the bees will know which of them—some of those in the swarm cluster—should be involved in choosing the swarm's new homesite. Thus it was that when I began writing my book *Honeybee Democracy*, I summarized the behavior of nest-site scouts as follows (on page 6): "Once bivouacked [in a beard-like cluster], the swarm will field several hundred house hunters to explore some 70 square kilometers (30 square miles) of the surrounding landscape for potential homesites, locate a dozen or more possibilities, evaluate each one with respect to the multiple criteria that define a bee's dream home, and democratically select a favorite one for their new domicile." After I had finished writing this book, however, I learned that nest-site scouts often start searching for a new homesite *even before* their colony has cast a swarm. Moreover, I learned that the nest-site scouts sometimes finish choosing the new homesite *even before* their colony has cast a swarm. When this happens, the swarming bees can do something that I find truly amazing: they skip the process of bivouacking on some structure near their old residence, and fly directly to the new dwelling place. It is rare for a swarm to make a straight shot from old homesite to new homesite, but it does happen. I will now describe the work that showed me that this can happen, and so revealed that nest-site scouts have even greater behavioral versatility than I realized when I wrote *Honeybee Democracy*.

The studies that I present in this chapter unfolded over the summers of 2007 and 2008. This is when Juliana Rangel—then a Ph.D. student at Cornell University; now the Professor of Apiculture at Texas A&M University—and I teamed up to investigate the mystery of what happens inside a hive to trigger the explosive departure of a swarm from its home. Anyone who has seen thousands of swarming bees pouring from a hive knows that it is an unforgettable sight. Where one minute you are watching foragers flying calmly to and from their hive, in the next minute you are engulfed by thousands of worker bees—and, somewhere among them, the mother queen—swirling wildly around the apiary. This buzzy commotion lasts only several minutes. Usually, it ends with the 10,000 or so bees assembling in the form of a beard-shaped cluster of bees that hangs in a bush or from a tree branch.

Juliana and I were itching to figure out what triggers a swarm's mass exodus from its home. We figured that we should start our investigation by watching and listening to what happens inside a hive shortly before a swarm bursts from its home. So, on 14 May 2007, we installed a strong, "queenright" colony—one with a healthy, mated, and egg-laying queen—in each of five observation hives. These hives were installed in the special room for observation hives at the Liddell Field Station of Cornell. Each hive had a tunnel that led outdoors, so the bees in these hives were able to come and go easily. Also, each hive had a small microphone mounted inside it so we could eavesdrop on the sounds made by the bees living in these hives before and during the process of swarm exodus. We covered the windows of the room housing the observation hives, so the light conditions inside each hive were like those inside a natural nest: dark everywhere except where daylight shone into the entrance. Honey bees use the light coming in their nest's entrance to find their way outside.

Because our study colonies were housed in observation hives, we were able to video record the bees' behavior during the process of swarm exodus. We could do so, even though our observation hives were in a dark room, by using the night-vision function (which uses infrared illumination) of our video cameras. Each camera's field of view was a 4-inch by 5-inch (10 × 13-centimeter) area of comb near the entrance of one of our hives. We then waited for the colonies to expand their populations and cast their swarms. Three of the five colonies did so, about three weeks after we installed these colonies in our observation hives.

What did we see and what did we hear? In the three colonies that swarmed, we found that a consistent set of events started up about 90 minutes before the swarming bees began to dash outside (Fig. 9.1, top). First, several dozen bees (hereafter called "activator bees") began to walk across the combs while producing worker piping signals. You may recall from Chapter 6 that Jürgen Tautz and I reported in 2001 that when a nest-site scout produces piping signals in a bivouacked swarm, she grabs other bees one at a time, presses her thorax onto each grabbed bee, and then activates her wing muscles (without her wings being spread) for about one second (see Fig. 6.4). This produces a body vibration of about 250 Hz that passes directly into the other bee's body and that, to the human ear, sounds like a shrill whine. Jürgen and I also reported in our 2001 paper that the piping signal stimulates the bees in a bivouacked swarm to warm up their flight muscles in preparation for the flight to their new home. What Juliana and I discovered in 2007 is that the piping signal is also produced by worker bees when their colony prepares to cast a swarm. Here, too, the piping signal stimulates bees to warm up their flight muscles, but now they do so in preparation for their exodus from their old home rather than their flight to their new home.

A second discovery that Juliana and I made in 2007 is that a few minutes before a swarm bursts from its home, many of the bees that have been producing piping signals switch to producing buzz-run signals (Fig. 9.1, bottom). In Chapter 7, we saw how Clare Rittschof and I reported in 2006 that when a nest-site scout produces the buzz-run signal in a bivouacked swarm, she runs over and butts into other bees while buzzing her wings. This behavior is depicted in Figure 9.2. Our 2006 paper reported that this buzz-run behavior is the "Let's go!" signal that

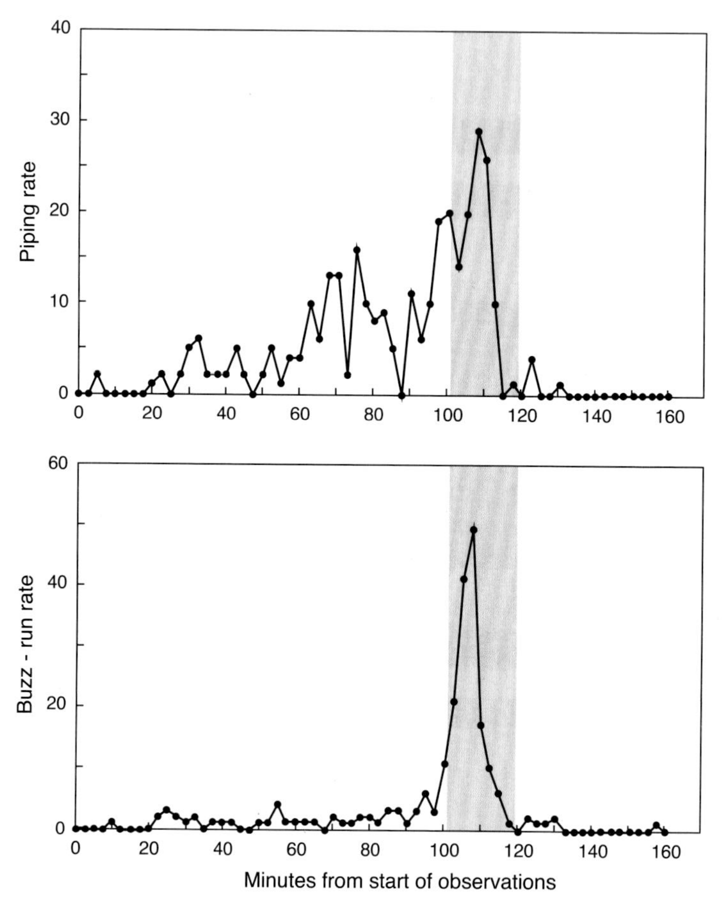

FIG. 9.1. Production rates of piping signals and buzz-run signals inside the nest of a colony before, during (tan boxes), and after a swarm's exodus. Numbers on vertical axes denote number of signals heard (piping signals) or seen (buzz-runs) near the hive's entrance in 30 seconds. This figure shows that piping signals are produced for an hour or so before a swarm leaves it hive, and that buzz-runs are produced only shortly before and during a swarm's departure.

stimulates everybody in the swarm to launch into flight for the journey to their new homesite. Now, in 2007, Juliana and I discovered that worker honey bees use the buzz-run signal not just to trigger the takeoff of swarm bees from their interim site, but also to stimulate the exodus of swarm bees from their home.

Having learned that it is the *same signals*—worker piping and buzz-runs—that initiate both a swarm's exodus from its home and a swarm's

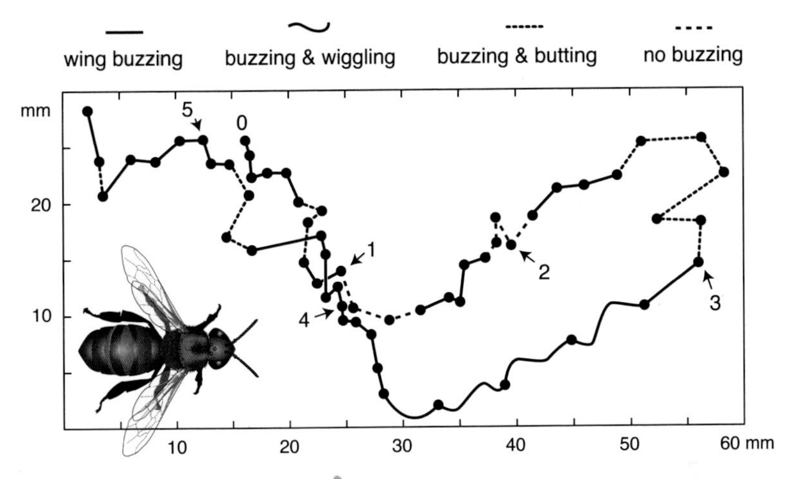

FIG. 9.2. Movement pattern of a worker bee producing the buzz-run signal over a time period of slightly less than 6 seconds. The numbers and dots along her track denote 1-second and 0.1-second intervals, respectively. 50 millimeters = 2 inches.

takeoff from its cluster site, Juliana and I wondered: Is it *the same bees*—the nest-site scouts—that trigger both departures? To answer this question, Juliana and I, together with Sean R. Griffin (then an undergraduate student at Cornell, now the Director of Science and Conservation at the Lady Bird Johnson Wildflower Center in Texas), conducted a further study in 2008. When we began this study, our goal was to answer just one question—*Is it nest-site scouts that trigger the mass exodus of a swarm from its home?*—but by the end of our study we had learned from the bees much more than just the answer to this particular question.

Let's turn now to what Juliana, Sean, and I did, saw, and learned in the summer of 2008. The steps of our investigation were as follows:

(1) We moved three colonies that we knew were preparing to cast a swarm to a remote location with few natural nest sites: Appledore Island, Maine. We knew that these three colonies were preparing to swarm because each one contained "swarm cells." These are the large, peanut-shaped structures made of beeswax that shelter developing queen bees. One of these new queens will replace a colony's original queen after she has departed in a swarm.

FIG. 9.3. Sean Griffin labeling scout bees that were visiting the green nest box that sits inside the lean-to shelter. The site is 750 feet (230 meters) from the barn that sheltered the observation hives housing our study colonies. The Atlantic Ocean stretches to the east in the background. When an unmarked scout bee went in the box, Sean captured her when she came out by holding a small insect net over the entrance hole. Then he pinned her gently between folds of the netting and put a dot of yellow paint on her thorax. Finally, he released her and recorded her marking, to keep count of how many scout bees he had labeled. The scouts showed no signs of being disturbed by getting netted and labeled; upon release, they returned to inspecting the nest box.

(2) We installed these three colonies, and their combs with the queen cells, in observation hives that we set up inside a barn on the island.
(3) We placed an unoccupied nest box in a shelter about 750 feet (230 meters) from the barn (Fig. 9.3).
(4) Sean labeled many of the bees that visited the nest box *before* each colony cast its swarm. These labeled bees were all nest-site scouts. Each one received a dot of yellow paint atop her thorax.
(5) Juliana and I monitored the behavior of the nest-site scouts (i.e., the labeled bees) when they came home and joined the crowd of

FIG. 9.4. Part of a cluster of bees on the side of the barn. The dark hole in this cluster is the entrance to the observation hive (inside the barn) which was the home of these bees. It was a hot and sunny day, so it was very hot inside the barn and many of the workers in each colony moved out of their hive and onto the side of the barn. This was helpful to us, because it made it easy to observe the scout bees (bearing yellow paint marks) and to track them with a microphone, to see which ones produced piping signals.

bees clustered around the small hole on the side of the barn that was the entrance to their home (Fig. 9.4). These clusters formed when many worker bees moved outside their colony's hive, to help prevent their colony from overheating. The weather was hot and sunny when we conducted this experiment, so every day the temperature inside the barn rose to over 90°F (32°C).

By labeling scout bees at the nest box, and then observing their activities back at their hive before their colony cast its swarm, we were able to address our main question: Is it nest-site scouts that trigger a swarm's departure from its home (by producing piping signals and buzz-runs)?

This study unfolded smoothly. What follows is a summary of the main events for our three colonies.

COLONY 1: On 1 July, scouts from this colony began to visit the nest box. Over the 3-day period 1–3 July, Sean labeled 462 scout bees there. At 12:52 p.m. on 3 July, this colony swarmed and the bees settled in a sumac tree (*Rhus glabra*) beside the barn. We then shook the swarm into a hive, and thereby cleared the nest box of scout bees from Colony 1.

COLONY 2: On 5 July, scouts from colony 2 began to visit the nest box. Over the period 5–7 July, Sean labeled 86 scout bees there. At 10:58 a.m. on 7 July, this colony swarmed and—*to our amazement*—the bees flew directly to our nest box! Around noon on this day, we removed the swarm bees from the nest box by shaking them into an empty hive that we positioned 60 feet (18 meters) from the nest box. This cleared the nest box of the bees from Colony 2.

COLONY 3: On 8 July, scouts from Colony 3 began to visit the nest box. Over the period of 8–9 July, Sean labeled 572 scout bees there. On 9 July, at 1:45 p.m., this colony swarmed and the bees settled in a sumac tree beside the barn. We shook the swarm into a hive, and thereby concluded our investigation.

What did we learn about the identity of the bees that produced the piping signals and buzz-runs at each colony's home hive, and so triggered each swarm's departure from its hive? Was it the bees that Sean labeled as nest-site scouts that produced these signals? Yes, in each case, it was! Figure 9.5 shows an example of the evidence that we gathered. The three panels in this figure show for each type of signal that we monitored—waggle dance, piping signal, and buzz-run—the proportion of the bees producing this signal at Colony 1's hive that bore a dot of yellow paint. These paint marks showed us which bees had scouted the nest box. Nearly

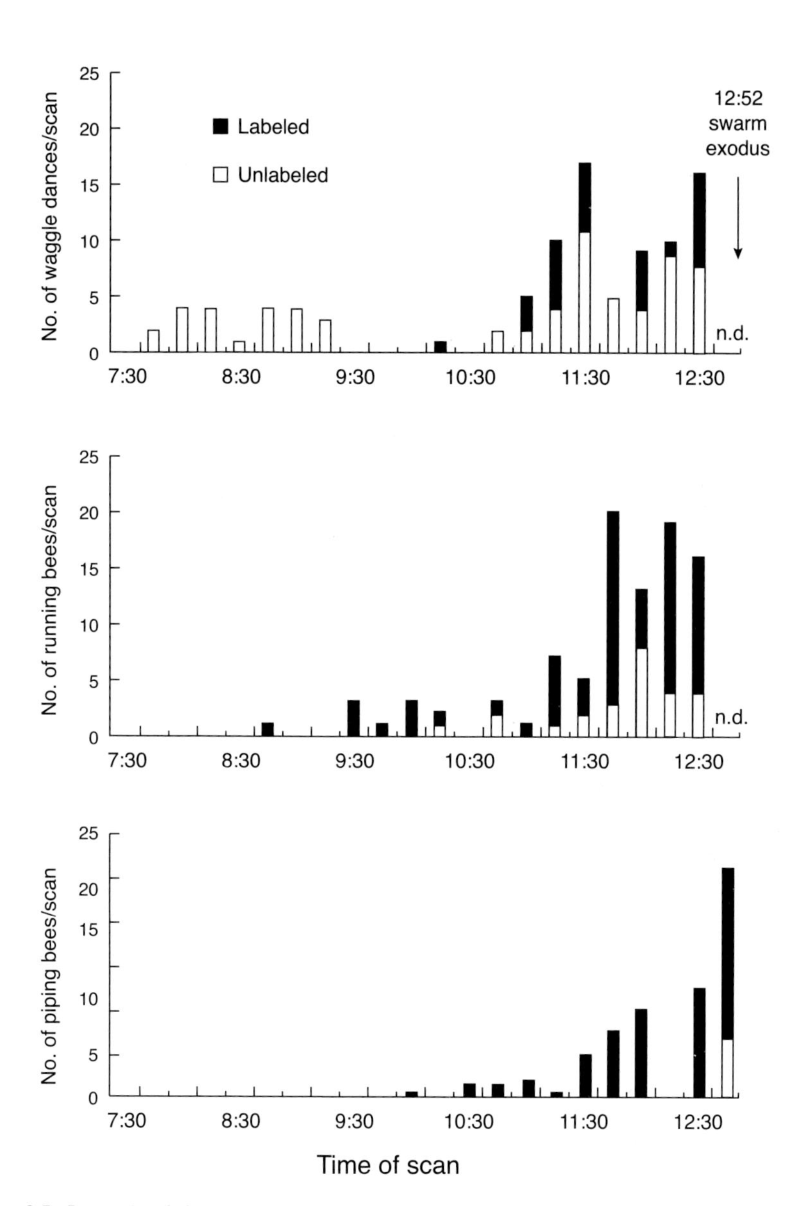

FIG. 9.5. Records of the production of signals at the hive of Colony 1, on the morning before it cast a swarm at 12:52 p.m. The counts for each signal are the number of bees observed producing the signal during a 30-sec scan at the hive. *Top*: Waggle dancers. *Middle:* Buzz-runners. *Bottom*: Pipers. Each bar in these plots shows how many bees were labeled with a paint dot (hence were known to be nest-site scouts) or were unlabeled. This figure shows that in the hour before the swarm's exodus from its hive, nearly every bee that was a buzz-runner or was a piper was labeled with a paint mark which indicated that she was a nest-site scout.

every bee that produced the piping signals and the buzz-runs bore a dot of yellow paint; this showed us that the "activator bees" in this colony were indeed nest-site scouts. (*Technical note: It was impossible for Sean to capture and put a paint dot on every scout bee that visited the nest box, so it is not surprising that some buzz-runners and pipers without paint marks were spotted at the hive of* Colony 1.)

This study on Appledore Island revealed much about the behavioral sophistication of the worker bees that function as nest-site scouts. *First*, it showed us that nest-site scouts can spring into action several days before their colony casts a swarm. *Second*, it showed us that sometimes the nest-site scouts will complete the nest-site selection process even before their colony casts its swarm. *Third*, it showed us that it is nest-site scouts that trigger the departure of a swarm from its home, by producing piping signals and buzz-run signals. And *fourth*, it showed us that if the nest-site scouts finish choosing a new homesite before their colony has cast its swarm, then when the swarm is cast it can fly directly to its new home, guided by the numerous nest-site scouts that know its location.

I have a special fondness for this small study because in addition to answering the question that was its focus—*Is it nest-site scouts that trigger the mass exodus of a swarm from its home*?—it deepened my respect for the behavioral versatility of worker honey bees. I had long known that the behavior of worker bees is amazingly complex. After all, their ability to share information about the locations of rich food sources (and good nest sites) by means of waggle dances is, by itself, almost fantastic. But I must say that it is the behavioral complexity of worker bees when they are functioning as nest-site scouts that evokes my most profound sense of admiration for these bees. I hope that you, too, admire deeply these marvelous little creatures.

CHAPTER 10

Messenger Bees

There is just one individual in a honey bee colony whose presence is critical. This is, of course, the queen bee. She is unique in being the longest-bodied and longest-lived member of a colony, but what really makes her special is her fertility. She is the mother of everyone else, and this "everyone else" can be tens of thousands of bees. About 95 percent of her offspring are workers—the small daughters that never mate and almost never have offspring, but perform all the tasks that keep a colony humming. The other 5 percent of a queen's offspring are her large daughters (queens) and her hulking sons (drones). These queen daughters and drone sons are the means by which a honey bee colony passes its genes on to future generations.

Because a colony has just one queen, it is essential that the nurse bees in a colony—the young worker bees that are responsible for tending a colony's brood—closely monitor the presence of "Her Majesty." She will die at some point, and when she does, the nurse bees must respond quickly—within about 24 hours of her death—to begin rearing her successor. If all goes well, this emergency queen rearing will produce a vigorous new queen who will mate with 10–20 drones from other colonies, and then will start her lifelong job of being the colony's egg-layer-in-chief. Thus the colony will live on, though with a somewhat different genetic makeup than before.

How do the nurse bees in a colony closely monitor their queen's presence, so they can respond quickly to her loss if she should die? It has been known since the 1950s that so long as a queen is alive, she signals her presence to her colony's nurse bees by secreting an oily material called "queen substance." Its primary constituent is the chemical compound that organic chemists call (E)-9-oxo-2-decenoic acid and that I will refer to by its nickname: 9-ODA. This substance is produced in a pair of large glands called the mandibular glands. They are so named because they are located in a queen's head, directly above her mandibles (jaws), as shown in Figure 10.1. We know that 9-ODA is slightly volatile, so it can disperse from a queen's body as a scent. This is how it functions as the sex attractant pheromone used by queens on their mating flights, and as the signal of a queen's presence in an airborne swarm (as was discussed in Chapter 8). But, as will be explained shortly, several studies have shown that smelling the low background level of 9-ODA in the air inside a hive *is not* how the young worker bees (a.k.a. the nurse bees) in a colony monitor their queen's presence. This tells us that a honey bee colony must have a special means of mass communication that, in human terms, is akin to one woman signaling her presence to some 10,000 people in a stadium at night with all the lights turned off. Now, imagine that this woman, like a queen bee, signals her presence with a scent that is registered only by individuals who are close to a source of the scent, say, less than an arm's length away. How does this mass communication between one signal sender (the queen) and thousands of signal receivers (the nurse bees) work? This is the mystery that I investigated in the summer of 1977 and that we will examine now.

My bee-biologist predecessors had made many discoveries that defined the puzzle of how the nurse bees in a colony are able to sense whether (or not) their mother queen is present. I found especially helpful the studies of the British bee researcher Dr. Colin G. Butler. He served from 1944 to 1972 as Head of the Bee Department at the Rothamsted Experimental Station, in the south of England. In the 1950s, he led a team of entomologists and chemists who identified the primary component of

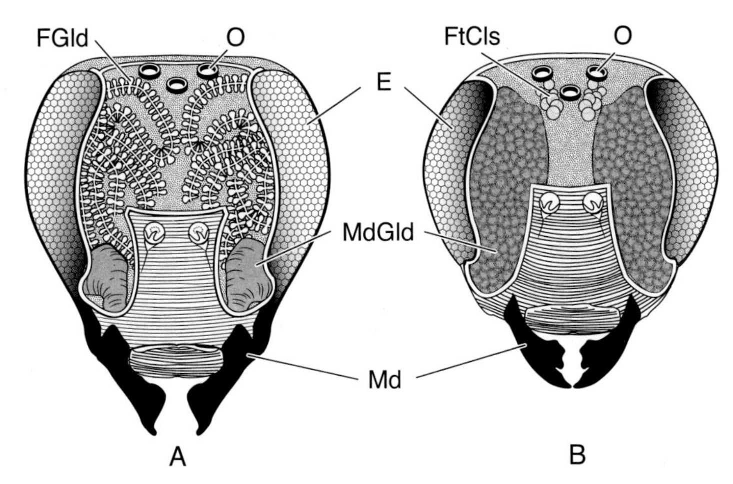

FIG. 10.1. Head of a worker bee (A) and a queen bee (B), with part of the facial area removed to show the difference in size of the mandibular glands (MdGld) in these two types of honey bee. The largest glands in the worker bee's head are the brood food glands (FGld), also known as the hypopharyngeal glands, but in the queen bee's head it is the mandibular glands that are the largest. Also depicted are the simple eyes called ocelli (O), fat cells (FtCls), compound eyes (E), and mandibles (Md). Note that the head of a queen is slightly smaller than that of a worker.

queen substance (9-ODA). Butler also determined that the workers in a colony must be able to *contact the queen* for there to be strong, colony-wide inhibition of queen rearing by the nurse bees. This finding showed that airborne dispersal of 9-ODA is *not* how this signal of the queen's presence is transmitted from queen to nurse bees inside a colony's nest. Furthermore, Butler found that the queen's signal of her presence fades rapidly when she is plucked from her colony. Within about 10 hours of losing their queen, the nurse bees will start to rear several replacement queens. This is called "emergency queen rearing." The fact that it begins soon after a queen is removed shows that the nurse bees must normally receive signals of their queen's presence every few hours or so. This puzzled me, because I knew that a honey bee colony contains many thousands of nurse bees. How does a queen's 9-ODA signal reach most, or all, of a colony's nurse bees every several hours?

After reading closely the papers of Butler and his colleagues, I figured that there were just two possible ways for a queen's 9-ODA signal of her

FIG. 10.2. A "retinue" of workers has formed around their queen, who is standing still. Besides feeding the queen, these workers are grooming her, licking her, and stroking her with their antennae. In doing so, they acquire traces of her queen substance (9-ODA) pheromone.

presence to be transmitted widely and pretty quickly to the nurse bees. The first is by *direct* (queen-to-nurse bee) contacts. It was known that a queen moves about her colony's brood nest and that several nurse bees can contact her simultaneously when she stands still (Fig. 10.2). It was also known that the turnover rate of the nurse bees that contact a stationary queen is high. It seemed possible, therefore, that a high percentage of a queenright colony's nurse bees could contact their queen directly, and that perhaps this was how the signal of her presence was maintained. But, I wondered, is this the full story? A second possible way for a queen's 9-ODA signal to be transmitted widely and quickly to the nurse bees is by nurse bees acquiring some of the queen's 9-ODA scent and then walking around to spread this chemical signal. If so, then the *direct* (queen-to-nurse-bee) transmission of the signal of the queen's presence might be supplemented by *indirect* (queen-to-nurse-bee-to-nurse-bee) transmission of this signal.

When I started my work on queen-to-nurse bee signaling, back in 1977, there was already good evidence that the indirect transmission mechanism is indeed part of the story. In 1959, a Dutch Ph.D. student at the University of Utrecht, Dr. Christina Verheijen-Voogd, had reported that workers that she plucked from a queen's retinue appeared to function as "substitute queens." Other workers were attracted to the plucked workers. This finding was confirmed in 1972 by another researcher at the University of Utrecht, Professor Hayo H. W. Velthuis. He reported that when he used clean forceps to pick up a worker bee that was in a queen's retinue and that was licking and touching the queen with her (the worker's) antennae, and then he put her in a cage with other workers, he saw that this worker attracted much attention from the other workers in the cage. Furthermore, both Dutch investigators found that when they put a retinue worker into a group of young workers from a *queenless* colony, the retinue worker attracted very close attention from the workers. Indeed, the queenless workers behaved toward the introduced retinue worker much as they would toward a queen. These studies were very helpful, because they showed that a worker bee standing near, and contacting, a queen can adsorb—probably in the waxy cuticle layer that covers her body—the chemical signal of a queen's presence. At this point, I asked myself, "Do the nurse bees that have had close contact with their queen walk around and help spread the signal of their queen's presence?" In other words, do they function as "messengers" that carry the signal of the queen's presence to their fellow nurse bees?

Both Christina Verheijen-Voogd and Hayo Velthuis conducted their studies with small groups of workers, just 60–65 individuals at most. This meant that it was not clear how their findings applied to the inner workings of full-size colonies. To avoid this shortcoming, I built at the Dyce Laboratory a large (41″/104 cm wide and 38″ /96 cm tall) observation hive. Its volume was the size of a natural nest cavity: 1.3 cubic feet (37 liters) (Fig. 10.3). The four wide frames in this hive held beeswax combs.

In early June 1977, I stocked my observation hive with approximately 15,000 worker bees and a queen. After leaving the colony alone for four

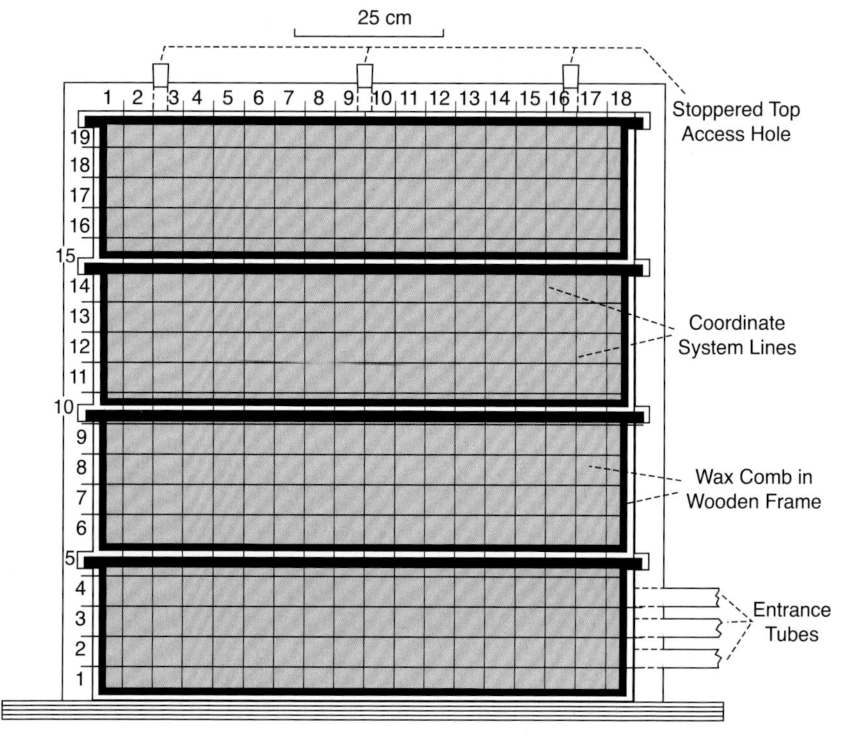

FIG. 10.3. The large observation hive used to study the behaviors of the workers that help a colony's nurse bees stay informed about their queen's presence by sensing her mandibular pheromone (9-ODA). The amount of comb in each frame was equal to what is found in two deep frames in a Langstroth hive.

weeks, to give it time to establish a brood nest and build up stores of honey and pollen, I added to it a cohort of 2,000 bees that were newly emerged (hence were 0 days old) and were individually identifiable. Each bee had been labeled with a colored and numbered disk on the thorax and a paint mark on the abdomen, like what is shown in Figure 4.3. Any labeled bee that died before she was 16 days old (i.e., before she started working outside the hive) was dragged out of the hive and dropped into a dead-bee trap at the hive's entrance. I recorded each day which of my labeled bees had died. So, I always knew how many of my labeled worker bees were still present in the colony on each day of my study, which continued until the labeled bees were 16 days old. I stopped then because in summer, very few worker bees are still functioning as nurse bees when they reach the age of 16 days old.

I made use of this cohort of same-age, individually identifiable bees in several ways. *First*, I used it to determine the age range of the workers that tended the queen. I did so by watching the queen steadily for two hours every three days, and recording the IDs of the labeled bees that tended her, i.e., that approached her and contacted her steadily for at least 10 seconds. This revealed that the labeled bees were likely to attend the queen when they were 4–10 days old, but not when they were younger or older. In other words, this showed that most of the prolonged contacts between the queen and workers were made by nurse-bee-age workers. *Second*, I used this cohort of individually identifiable bees to determine what fraction of them (when they were 4–6 days old, thus were functioning as nurse bees) made direct contact with their queen. I found that within a 5-hour period, 20 percent of these bees did so, and that within a 10-hour period, 35 percent did so. This told me that fewer than half the nurse bees in this colony were making direct contact with their queen within a 10-hour period. *Third*, I observed and recorded—by tape recording what I said aloud about what I was seeing—how the workers that were 4–10 days old behaved when they contacted their queen while she was standing still. Many of them did so with what looked like hesitation; they would touch her lightly with their antennae and then quickly withdraw by stepping back. Others, however, contacted her closely and persistently. Sometimes they swiped their antennae over her in a flurry that lasted about 30 seconds, and other times they licked her with their tongues for about 10 seconds. After doing these things, these bees usually paused to groom their antennae and tongue with their forelegs. These observations showed that if a nurse bee-age worker picks up 9-ODA when she contacts her queen, then it gets smeared over her (the nurse bee's) forelegs, tongue, and antennae.

The most intriguing thing that I saw, though, was what the nurse bees that had prolonged contacts with the queen (a.k.a. the "retinue bees") did after they had backed away from her. They first groomed their antennae and mouthparts for several seconds, then they walked rapidly and widely around the brood-nest region in the hive for several minutes. In contrast to the retinue bees, the control bees did *not* groom themselves extensively, and they did *not* walk excitedly around in the brood nest (Fig. 10.4). On average, the retinue workers walked through 59 of the grid squares on the glass walls of the observation hive, but the control bees walked through just 26 of these grid squares. It was clear that the retinue workers had become excited by "meeting" their queen. (*Technical note: I recorded the movements of ten retinue bees, one at a time, after each one had stepped away from the queen. I did so by locking my eyes on an ex-retinue bee and then continuously recording for 30 minutes her location (using a wax pencil) on two thin sheets of glass that I had placed over the thick window-walls of my observation hive. I used the same methods to record the movements and activities of ten "control bees," i.e., nurse bees that were working far from the queen.*)

Having seen the large difference in movement levels between retinue and control workers, I wanted to compare their behaviors in finer detail. To do so, I chose the bees as described above, but now I wore a headset of magnifying lenses so I could watch closely the behaviors of each focal bee, i.e., the bee that I was watching. Again, I followed each worker bee (retinue or control) for 30 minutes, and I tape-recorded my running narrative of what this bee did and how she was treated by other bees nearby. It would have been better to video record everything and then analyze the video recordings in detail, but I conducted this study years before I had video equipment.

This was eye-straining and mind-straining work, so I managed to watch only 10 pairs of nurse bees (retinue and control) over three days. But this work yielded a discovery. It revealed conspicuous differences in how the retinue and control bees behaved and in how they were treated by their fellow worker bees. Specifically, when I compared the two groups in terms of (1) frequency of being inspected by nestmates, (2) number of bouts of "antennation" (mutual antennal contacts) with nestmates, and (3) number of food exchanges with nestmates during the 30-minute

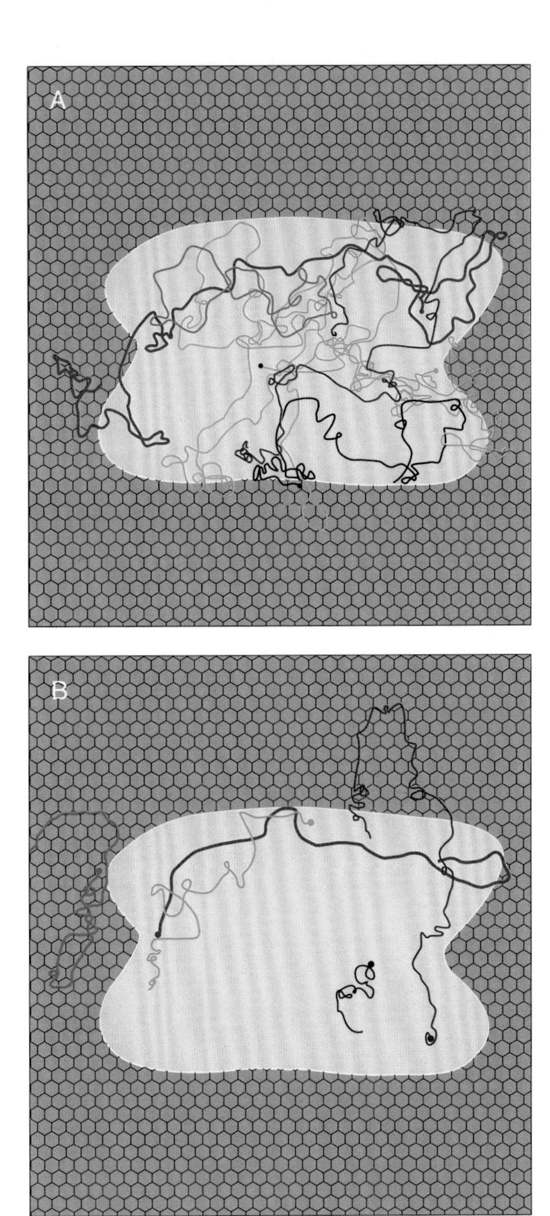

FIG. 10.4. Records of the movements of individual nurse bees for 30 minutes. **A.** Tracings of retinue bees after breaking contact with the queen. **B.** Tracings of control bees. Dots denote starting points. Some individuals walked around on both sides of the comb; their movements have been combined into a single, winding line. Tan area indicates comb that contained brood; brown area represents comb that contained food or was empty. These figures show that nurse bees that have contacted their queen, relative to ones that have not, move more quickly and widely around the brood nest.

periods when I traced their movements, I found that the retinue bees, relative to the control bees, received more antennal inspections from nestmates (14 vs. 1) and experienced more mutual antennations (56 vs. 12). Moreover, I found that among the retinue bees there was a strong correlation ($r=0.76$) between the duration of a bee's contact with the queen and the number of times she was inspected by nestmates over the next 30 minutes. But retinue bees did not conduct more food exchanges than control bees (1.8 vs. 1.7). And, relative to the control bees, the retinue bees performed fewer than one-third as many "labor acts," e.g., cleaning a cell, feeding a larva, fanning for nest ventilation, dragging a dead bee, trimming scraps of old lids from empty brood cells, and shaping bits of comb. The retinue bees appeared to be focused on moving around the brood nest and engaging in mutual antennal contacts with their hive mates.

All these findings support the idea that nurse bees that have had thorough contacts with their queen pick up 9-ODA on their bodies and then disperse this signal of the queen's presence to their fellow nurse bees by walking around the brood nest and having bouts of antennation with other nurse bees.

This work led me to coin the term "messenger bees" to refer to nurse-age bees that make extensive contact with their queen and then move widely around their colony's brood nest. The role of a messenger bee lasts only briefly, certainly less than 30 minutes. An example of this is shown in Figure 10.5, which shows the record of one messenger bee's behavior over the 30-minute period—from 10:30 to 11:00 in the morning—immediately after she stepped back from the queen. We see that her antennations with other bees in the brood nest are concentrated in the first half of this 30-minute period, and that her bouts of working (cell cleaning, feeding a larva, fanning, dragging a dead bee, shaping bits of comb, and trimming the wax cappings of cells) appear entirely in the second half. This pattern of frequent antennations with others in the first 15 minutes and then working on her own in the second 15 minutes was

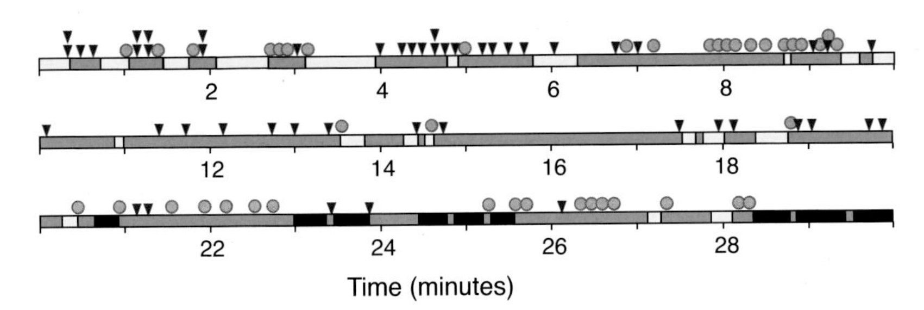

FIG. 10.5. Time line that shows the record of one messenger bee's behavior during the 30 minutes after she broke contact with the queen. During the first 20 minutes, she focused on walking around the brood nest and having antennal contacts ("antennations") with other workers.

found in the records of all 10 retinue bees, but it was absent from the records of all 10 control bees.

Next, I attempted to obtain direct evidence that messenger bees acquire the queen's mandibular pheromone (9-ODA) by licking and antennating. My approach was straightforward: compare the levels of 9-ODA in ethanol extracts of the bodies of 250 messenger bees and the bodies of 250 randomly chosen bees in the center of a colony. But I detected no 9-ODA in either of my extracts. Looking back, I am pretty sure that the detector in the late 1960s gas chromatograph that I used was not sufficiently sensitive for this analysis.

It was wonderful, therefore, when two biologists (Ph.D. student Ken Naumann and Professor Mark L. Winston) and an organic chemist (Professor Keith N. Slessor) at Simon Fraser University in Canada, plus two natural products chemists (Professors Glenn D. Prestwich and Francis X. Webster) at the State University of New York, teamed up in the late 1980s to study the production and transmission of 9-ODA in honey bee colonies. They brought to their investigation a powerful tool for studying the movement of 9-ODA within a colony: a version of 9-ODA that was made radioactive by replacing some of its hydrogen atoms with a radioactive isotope of hydrogen called "tritium." They applied their radioactively labeled 9-ODA to a lure (a dead worker bee scented like a queen bee) and then they put it, and 15 worker bees, in a Petri dish. After they allowed the workers to contact the lure for fixed periods (0, 5, 30, or 60 seconds), they removed the workers and measured the levels of radioactivity on the various parts of their bodies. These body parts included the workers' antennae, mouthparts, legs, thorax, and abdomen. This team of investigators also developed a method for measuring a queen's production of 9-ODA in her mandibular glands.

The research methods developed by this team of biologists and chemists produced beautiful results. They discovered that a queen releases about 138 micrograms of 9-ODA daily, and that almost all of it (ca. 135 micrograms) is acquired and dispersed by the messenger bees. Only about 3 micrograms ends up in the wax combs. This team also learned that it is the messenger bees that *lick the queen* that acquire most of her 9-ODA (121 micrograms); those that only *antennate the queen* acquire only a small portion of her 9-ODA (14 micrograms). They also found that after a messenger bee has acquired a quantity of 9-ODA on her tongue, she transfers it to her abdomen. She does so by grooming her tongue with her forelegs and then passing the oily 9-ODA rearward. Presumably, this makes the 9-ODA that she has picked up more conspicuous to other workers than if she were to leave it on her mouthparts.

These investigators also discovered why the 9-ODA in a colony does not build up over time and potentially create a false signal of the queen's presence: the workers ingest it and they absorb it in the wax layer that covers their bodies. Then the 9-ODA is, somehow, broken down rapidly. Its half-life on the surface of a worker is only about 13 minutes. This speedy "decay" of a queen's signal means that the workers are able to sense quickly when their queen has died. This quick sensing is critical to the future of a colony. Why? Because the workers need female larvae that are less than one day old in order to rear a batch of high-quality queens, one of whom will become the dead queen's successor. Unlike in our own societies, which have Vice-Presidents, Vice-Chairs, and such to cope with losses of leaders, honey bee colonies do not have Vice-Queens. Instead, honey bees have the queen substance/messenger bee system so a colony can sense quickly when its mother queen has died and thus when it must kick-start its emergency queen rearing to produce her successor.

CHAPTER 11

A Tale of Four Species

In the spring of 1977, I began to see that I would finish my Ph.D. studies in about a year, so I started to ponder what I should study next, when I would be a postdoctoral student. Another field-based study of honey bee behavior? A switch to laboratory-based studies of their chemical communication? Something else? I made a list of potential projects and pondered the pros and cons of each. This exercise made it crystal clear that what I hoped most to do next was a field study of three Asian species of honey bees—*Apis florea*, *Apis cerana*, and *Apis dorsata*—living in the wild in Thailand. I was keen to study these bees partly because there was much about their lives that remained mysterious, and partly because I wanted to compare their ways of life to that of the wild colonies of *Apis mellifera* that I had studied for my Ph.D. thesis. My goal would be to understand better the adaptive "design" of honey bee colonies, by looking at four species in the genus *Apis*: *A. mellifera*, *A. florea*, *A. cerana*, and *A. dorsata*.

It was known that these four species *share* many remarkable traits: all have colonies with a single queen and thousands of workers, all build vertical combs of hexagonal cells made of beeswax, all have a division of labor by age among the workers, all reproduce by swarming, and all possess the waggle dance. It was also known, however, that these four species *differ* in many conspicuous traits: worker body size, colony size, nesting behavior, and defensiveness (Fig. 11.1). I figured that a comparative study of how the colonies of each species live in nature could help biologists understand why their colonies are so different, even though these four species are all descendants of one ancestral species that lived in the Asian tropics approximately 70 million years ago. Fossil bees found in amber from this time period (geologists call it the Late Cretaceous) show us that the ancestor of all the honey bee species now in existence was a truly social ("eusocial") bee. In other words, the ancestor to all the honey bees we see today possessed three special features: (1) it lived in groups that shared a nest, (2) there was a dominant egg layer in each group, and (3) the dominant egg layer (the queen) was assisted by her daughters (the workers). The fossil record also shows us that the workers of the ancestral honey bee species had pollen baskets on their hind legs.

Assuming that somehow or other I would find a way to get to Thailand and study the Asian honey bees, I developed a general plan of study. The first goal would be to describe *how* the colonies of each species live in nature. This would involve finding a study site where wild colonies of all three species live, locating the nests of ten or more colonies of each species, and then spending countless hours observing these colonies at all times of day and night, taking notes, making drawings, and getting measurements. I knew that this would give us a much fuller picture of the natural history of each species. The second goal would be to understand *why* the colonies of these three species differ in so many ways, including worker size, nest site, comb structure, colony size, and colony defensiveness. In other words, I would try to identify each species' "strategy" for colony survival and reproduction. This part of the investigation might require experimental work to test the adaptiveness of certain colony-level traits, such as nest visibility and nest structure.

One of the reasons I was strongly attracted to studying how colonies of the three Asian species of honey bee live in their tropical homelands was that I knew that this work would deepen my appreciation of the suite of adaptations that colonies of *Apis mellifera* possess to survive in cold-temperate regions of the world, specifically, northern Europe and North

FIG. 11.1. Workers of the three species of honey bees native to Thailand. Note: a worker of *Apis cerana* is about the same size as a worker of *Apis mellifera*. Artist: Sandra Bidwell Olenik.

America. I took inspiration from Charles Darwin's comparative studies of species that are closely related evolutionarily but have diverged ecologically, morphologically, and behaviorally—e.g., blind *cave* beetles vs. sighted *field* beetles. I was also inspired, and even more strongly, by the work of a Swiss biologist, Esther Cullen. She had done a beautiful comparative study of two closely related species of seabirds, the black-legged kittiwake (*Rissa tridactyla*) and the herring gull (*Larus argentatus*). For this project, she went to the Farne Islands (off the east coast of England), where individuals of both species breed, and she patiently watched them throughout their breeding seasons. This work led to the insight that the many distinctive ways that kittiwakes differ behaviorally from their sea gull relatives—alarm calls are barely audible instead of astonishingly loud; nest is a deep cup built of seaweed and mud instead of a shallow depression scraped in soil; young stay on nest instead of leaving shortly after hatching; tameness to human observers instead of fierce defensiveness; and so forth—are a suite of adaptations to the kittiwakes' unusual nesting habitat: narrow ledges on seaside cliffs. This nesting habitat differs markedly from that favored by sea gulls: grassy areas on sand dunes.

Once I had my dream research project clear in my mind, I began exploring ways to fund it. One possibility was to get a grant from the National Geographic Society; this could cover the travel and living expenses for fieldwork in South Asia. Another possibility was to be elected a Junior Fellow in the Society of Fellows at Harvard University. Each autumn, the Senior Fellows of this Society interview a few dozen candidates from around the world and then select eight of them to be Junior Fellows. Those chosen are offered three years of generous financial support. My PhD advisors, Bert Hölldobler and Edward O. Wilson, nominated me for one of these prize fellowships, and in August 1977 I was invited to an interview with the Senior Fellows, all highly distinguished members of the Harvard faculty, some even Nobel laureates. I was not sure that I would present myself well to the Senior Fellows, for I am by nature a shy person and I knew I would be extremely nervous. But I recalled reading in Karl von Frisch's autobiography how he had coped with his nervousness when, as a young student at the University of Munich, he had to summarize in his zoology class a scientific paper on the acrobatics a starfish performs to right itself when it has fallen on its "back." His landlady made him a model starfish from a piece of colorful silk and some wadding, and he took it, tucked in his breast pocket, to his presentation. When he pulled out his floppy model to demonstrate a starfish's various tricks, its sudden production caused laughter, which boosted his self-confidence. So, I brought to my interview a small wooden box of pinned specimens of the Asian honey bees, to show the Senior Fellows the bees' striking differences in size: *Apis florea* is smaller than a housefly, *Apis dorsata* is bigger than a bald-faced hornet, and *Apis cerana* is in between. I kept the box closed until I was asked to explain my proposed plan of study. Its opening caused the Senior Fellows to smile, lean forward, and ask me about my "wares." I think this show-and-tell helped, for I was elected a Junior Fellow, an appointment that I held from July 1978 to June 1980.

From early October 1979 to early May 1980, my wife, Robin, and I conducted a comparative study of how the three Asian honey bees live in the forested mountains of northeast Thailand. Robin is also a field biologist, but she focuses on the seaweeds and creatures that live in the chilly world of rocky intertidal zones, not on creatures like honey bees that favor sunny stands of flowers. Our paths crossed on Appledore Island in August 1977. We married in April 1979 in preparation for a seven-month "honeymoon" studying honey bees in Thailand.

We worked in two locations: the rugged Khao Yai National Park and the nearby National Corn and Sorghum Research Center (Suwan Farm) of Kasetsart University. Both sites had forest-covered hills populated with wild colonies of the Asian honey bees. We received much help from Professor Pongthep Akratanakul of Kasetsart University and his graduate student, Amnat Banomayong. Pongthep and I had become friends at the Dyce Laboratory at Cornell, where he had studied for his Ph.D. with Doc Morse. Pongthep helped us settle in at Suwan Farm and solve logistical problems, such as buying a used car (from his sister). Back then, Pongthep was an assistant professor in the Entomology Department; now, he is the Research Director of the university's Center for Agricultural Biotechnology. This is a good example of the cream rising to the top. Amnat introduced Robin and me to Thai culture and helped us learn Thai. We, in turn, introduced Amnat to field biology and helped him polish his English. The four of us made a good team.

Forty years ago, few tourists trekked up to the vast Khao Yai National Park (area: 860 square miles/2,200 square kilometers) in the mountains of northeast Thailand, and it was a biologist's paradise. Now, two big highways stretch northward from Bangkok to this region, and more than a million people visit the Khao Yai [Big Mountain] National Park each year. Robin and I remember times in the forest, by ourselves, listening to the loud whooping calls of white-handed gibbons (*Hylobates lar*) and meeting wild Asian elephants (*Elephas maximus*) ambling along trails. Also, I will never forget the sight of great hornbills (*Buceros bicornis*) flying majestically past the exposed-comb nest of an *Apis dorsata* colony that hung from the bottom limb of a towering Tualang tree (*Koompassia excelsa*) in Khao Yai (Fig. 11.2). Just as intriguing, at least to me, were the flocks of red junglefowl (*Gallus gallus*, the ancestor of domesticated chickens) chugging across the jungle floor, each flock led by a brightly colored male.

Our plan of work was straightforward: conduct a comparative study of the natural history of the three species of honey bees—*A. florea*, *A.cerana*, and *A. dorsata*—that live wild in the Khao Yai National Park and the forested hills around Suwan Farm. Previous studies of these bees gave us a good "sketch" of each species' biology. Our goal was to produce a more detailed picture of how these three species differ in morphology, nesting habitat, nest structure, array of defenses, colony mobility, and other traits. So, we began this work with our eyes wide open. Our plan was to let ourselves be guided by whatever we saw. This initial block of descriptive work in 1979–1980 set the stage for a second block of analytic work that ran from December 1984 to September 1985. The second investigation was conducted mainly by a postdoctoral student, Dr. Fred C. Dyer, and an undergraduate student, Josh L. Schein. Again, Professor Pongthep Akratanakul provided scientific and logistical support. Now the work had a tighter focus than before, for now we appreciated more fully that the fivefold range in worker body mass among these three species has profound effects on the range of temperatures over which the worker bees in each species can fly. We expected that this, in turn, would affect strongly how the colonies of each species function.

Let's start to make our comparisons of the three honey bee species found in Asia, and the one honey bee species (*Apis mellifera*) native to

FIG. 11.2. A glimpse of the forest setting of our studies in the Khao Yai National Park in Thailand. Robin is watching a colony of *Apis dorsata* that has built its nest 62 feet (19 meters) off the ground, on the underside of the lowest limb of a dead dipterocarp tree that is standing about 450 feet (ca. 150 meters) away, on the side of a stream.

Europe, the Middle East, and Africa, by reviewing the distinctive traits that all four species share. These are the traits that show us that these four species all belong to the genus *Apis*, i.e., they are all true honey bees. *First*, their nests consist primarily of vertical combs made of wax secreted from glands on the undersides of the worker bees' abdomens. *Second*,

TABLE 11.1. Comparison of worker and colony traits in four species of *Apis*.

	Open-nesting species		*Cavity-nesting species*	
Property	A. florea	A. dorsata	A. mellifera	A. cerana
Body mass (mg)	23	118	77	44
Combs/nest	1	1	5–8	5–6
Curtain over comb(s)	yes	yes	no	no
Workers/comb cell	3.99	3.28	0.80	0.74
Trips/forager/day	ca. 5	ca. 1	10–15	ca. 20
Flight metabolic rate at 25°C (watts/kilogram)	400	358	643	700
Longevity in summer (days)	50+	no data	ca. 35 max	ca. 35 max

their combs consist of two layers of hexagonal cells opening in opposite directions. *Third*, the cells that they use for rearing brood and storing food (honey and pollen) are all one size, but the cells that they use for rearing drones are distinctly larger (except in *Apis dorsata*, the giant honey bee). *Fourth*, the cells that they use for producing queens are not in the combs and are not horizontal hexagons; instead, they hang from the combs and resemble peanut shells. *Fifth*, their colonies are perennial, i.e., they can survive winters and thus live for several years. *Sixth*, their way of forming new colonies is a process of colony fissioning in which the mother queen and a swarm of workers leave the original nest to build a new one in a different location. And *seventh*, their workers perform waggle dances to share information about the locations of nest sites and sources of food, water, and resins. The fact that all four species share these seven distinctive traits indicates that they are all derived from a common ancestral species. In other words, these species are members of a "monophyletic group." This is an important point, for it means that the differences in morphology and behavior that exist among these species are adaptations to four different ways of life.

To begin to understand how and why these four honey bee species are different, it helps to note that they sort out into two pairs, as shown in Table 11.1. In one pair, the species nest in the open and in the other pair,

they nest in cavities (Fig. 11.3). (Note: several studies tell us that building combs in the open is ancestral to building combs in cavities.) Specifically, we see that the two species with the smallest and the largest workers, *A. florea* and *A. dorsata,* build exposed nests that have just one comb, while the two species with intermediate-size workers, *A. cerana* and *A. mellifera,* build their nests by "squeezing" several combs inside cavities, usually caves and hollow trees. We see, too, that the combs of the open-nesting species, unlike those of the cavity-nesting species, are protected by thick "curtains" of interlinked worker bees. These curtains provide protection in several ways: insulating the brood against heat loss, shading it from direct sunlight, and protecting it from potential predators. One important consequence of using thick curtains of bees for nest protection is that the colonies of the open-nesting bees, *A. florea* and *A. dorsata,* have about five times more workers per comb cell than the colonies of the cavity-nesting bees, *A. mellifera* and *A. cerana.*

How do colonies of the open-nesting species produce such a large "excess" of workers relative to the cavity-nesting species? Do the open-nesting species have a far shorter development time from egg to adult, so each comb cell in their nests produces many more bees in a given period of time? The answer is no. The worker development time is similar for all four species: 18–22 days. This means that the striking difference in number of workers per comb cell exists because the workers of the open-nesting species have longer lives than the workers of the cavity-nesting

FIG. 11.3. Typical nests and nest sites of the three honey bee species studied in Thailand. *Top left*: Nest of *Apis florea*, about 8 inches (20 cm) wide, built on a slender branch of a small tree. The vegetation that originally hid this nest was removed to take this photo. In this species, and in *A. dorsata*, the single comb of a colony's nest is covered, and thereby protected, by a thick blanket of bees. *Bottom left*: Nest of *Apis cerana* exposed in a cut-open tree cavity. The bees have been removed to show the combs. The two taller combs are ca. 12 inches (30 cm) tall. *Top right*: Two nests of *Apis dorsata*, about 65 feet (20 meters) high on the undersides of limbs of a massive dipterocarp tree. Each nest is about 40 inches (1 meter) wide. *Bottom right*: Comparison of combs of *A. florea* and *A. dorsata*; the *dorsata* comb shown is only one half of a colony's full comb.

species. One sign of this difference in worker lifespan is the way that the workers of *A. florea* and *A. dorsata* have low-tempo lives (and so have a low turnover rate), whereas the workers of *A. mellifera* and *A. cerana* have high-tempo lives (and so have a high turnover rate). Table 11.1 shows that this difference in tempo of life shows up in two ways: fewer trips per forager per day and lower flight metabolic rates for workers of *A. florea* and *A. dorsata* relative to those of *A. mellifera* and *A. cerana*.

Another good way to compare the three Asian species is to look at how the colonies of each species deal with a common foe, the giant Asian hornet, *Vespa tropica*, which is also called the "greater banded hornet." For a colony of *Apis florea*, whose small workers are no match for these huge wasps, and whose nest is not sheltered inside a protective cavity, the only good defense appears to be concealment by nesting in dense vegetation. We scored the visibility of the nests of each honey bee species by counting the number of directions from which we could spot a nest (or its nest entrances). The maximum value was 6: 4 sides plus above and below—when we stood 6 feet (2 meters) away. We found that nearly every nest of *florea* scored only 0–1, but that almost every nest of *cerana* and *dorsata* scored 4–6. Also, we found that at the start of the dry season, when many plants shed their leaves, the *florea* colonies that lost their nest concealment abandoned their nests and built new ones in places with good cover. When we performed an experiment of removing the vegetation around the nest in one set of *A. florea* colonies, but leaving this vegetation intact in another set of these colonies, we found that the colonies of *A. florea* that lost their concealment absconded within 12–15 days (long enough for their pupal brood to emerge), whereas the colonies that did not lose their concealment did not abscond. We also found that after an *A. florea* colony that had been exposed—by our removal of the vegetation hiding its nest—moved to a new homesite, its workers came back to their old nest to collect wax for building their new nest (Fig. 11.4). They performed this salvage operation by biting off flakes of wax, chewing them, pressing them into the "pollen baskets" on their hind legs, and then flying to their new nest site. I admire the thriftiness of these bees.

FIG. 11.4. Workers of *Apis florea* salvaging wax from a nest that their colony abandoned when it became exposed. The bees vacated this nest when the shrub that had concealed it shed its leaves at the start of the dry season. The colony did not, however, leave its exposed nest immediately. It waited about two weeks, long enough to give most of its brood time to mature, then it moved about 300 feet (100 meters) to a new, and highly cryptic, nesting site.

Evidently, this is a trait that has been favored by natural selection because the colonies of these little bees are weakly defended. They cope with predator attacks by abandoning a nest when it becomes exposed, and recycling wax from their old home to build their new home.

The peril of poor nest concealment for a colony of *Apis florea* was made clear to Robin and me when we watched a colony of *A. florea* whose nest hung in the open from the underside of a palm frond being attacked by a team of giant Asian hornets (*Vespa tropica*). It was not a fair fight. Three giant hornets overwhelmed the defenses of this colony of some 6,000 dwarf honey bees. The first hornet landed about an inch from the bees' nest, and then lunged repeatedly toward the nest. Each time she did, she snagged a worker bee in her jaws, crushed it, and dropped it . . . all in the blink of an eye. This one hornet murdered 52 *A. florea* workers in the next 20 minutes. Soon several more hornets joined the attack. The next morning the bees abandoned their nest, whereupon several hornets came in, killed off the few dozen straggler worker bees—helpless young things not yet strong enough to fly away—and then proceeded to "strip mine" the abandoned comb for the nutritious bee larvae and pupae it held. Watching this destruction, we saw clearly why colonies of the dwarf honey bee, *Apis florea*, usually occupy cryptic nest sites, such as inside shrubs with dense foliage.

Colonies of *Apis cerana* and *Apis dorsata* are much less vulnerable to attack by these giant hornets. When these awesome predators attack a colony of *A. cerana*, the bees retreat into their nest cavity, and this forces the hornet to go inside to press her attack. But if she does, then she will become covered by a ball of worker bees feverishly contracting their flight muscles to produce heat. The temperature inside this ball will shoot up to over 113°F (45°C) . . . hot enough to kill a giant hornet! I know this works well because I kept two colonies of *A. cerana* in hives at the Suwan Farm, and when I inspected these colonies, I sometimes found a dead hornet or two inside the hives. The hornets had been killed by being "cooked."

Whenever Robin and I watched a giant Asian hornet approach the nest of a colony of *Apis dorsata*, we saw the bees in the outermost layer of the colony's protective curtain perform a "shimmering behavior." The curtain bees closest to the hornet would flip their abdomens outward, and this would trigger a wave of abdomen-flipping bees that swept across the curtain of bees protecting the colony's nest. It certainly looked like a warning: "Don't mess with us!"

Some weeks later, I did a risky experiment that involved donning a white bee suit, climbing 60 feet (20 meters) up to a colony of *Apis dorsata* that had built its nest on the underside of a concrete water-tower tank, and then swinging my binoculars case (made of black leather) past this colony. The instant I did this, over 100 bees burst from the nest and began attacking my binoculars case, me, and the Thai villagers below who were watching. I quickly climbed down and dashed away, clutching my binoculars case, which now reeked with alarm pheromone from dozens of stingers implanted in it. My plan was to end my attack and draw the bees' "sting fire" away from the villagers. This worked. That evening, I extracted 72 wickedly large stingers from my binoculars case. The explosive defense that I saw that afternoon gave me an unforgettable memory, and made crystal clear to me why this species of honey bee, *Apis dorsata*, has been described as "the most ferocious stinging insect on earth."

CHAPTER 12

Colonies Are Information Centers

I still remember what I saw the first time I lay on the grass beside a hive housing a strong colony, and watched its foragers flying off to work: countless bees crisscrossing the blue sky like shooting stars. The time was a late afternoon in September 1969, and the place was the small field behind my parents' house where I kept my first colony of honey bees. Seeing those bees zooming away filled me with wonder. Where were they going? To the abandoned fields to the north, where stands of goldenrod (*Solidago* spp.) were blooming among thickets of gray dogwood (*Cornus racemosa*) and groves of quaking aspen (*Populus tremuloides*)? Perhaps to the west, to flower gardens at Cornell University? Maybe to the east, to other abandoned fields in the far end of Ellis Hollow? To the south, to the dairy farm atop Snyder Hill, where the alfalfa and white clover in the hayfields were still in bloom? Or to all of these places, and still more? What I saw on this afternoon became a lasting memory and inspired a long string of investigations into the genius of honey bee colonies at conducting search-and-retrieval operations. Set a hive of honey bees in your backyard, and within an hour foragers will be streaming into it with loads of pollen and nectar.

In June 1979, I returned to Ithaca to work with a friend, Kirk Visscher, who was starting his studies as a graduate student in the entomology department at Cornell. Roger A. Morse ("Doc") was his thesis advisor. Kirk and I had met at Harvard University in 1977, when Kirk was an undergraduate student concentrating in biology. We became friends when I supervised Kirk's honors thesis project on the mechanisms whereby dead worker bees are recognized as such and then are quickly removed from their colony's nest. Kirk and I decided to team up in the summer of 1979 because we both needed to start a research project, and we figured it made sense to work together. Kirk wanted to find a good project for his master's thesis, and I wanted to conduct the first step of a long-term project that I had in mind. I was thinking long-term because a few months before I had accepted an offer to become an assistant professor in the Department of Biology at Yale University. This meant that now I had the opportunity and the wherewithal to explore in depth some part of the inner workings of honey bee colonies.

Yale was a fine setting in which to develop further as a teacher and researcher. My teaching assignment was an undergraduate course titled Ethology (a.k.a. animal behavior), a subject that matched my interests. Also, the university was renovating the old caretaker's house in the Othniel C. Marsh Botanical Garden, converting the first-floor rooms into a honey bee laboratory and the upstairs rooms into a cozy apartment for Robin and me. This meant that we would be able to work and live in a park-like setting within the city of New Haven, Connecticut. All of these good things came, of course, with high expectations for my performance, especially as a researcher. So I needed to think carefully about my program of future studies with the bees.

Bernd Heinrich, an insect physiologist at the University of California, Berkeley, had recently written a book, *Bumblebee Economics*, which showed beautifully the value of viewing a bumble bee colony as a functional unit shaped by natural selection to be efficient in the collection and consumption of its energy supplies. This perspective certainly seemed right for honey bees, too. Moreover, because honey bee colonies possess communication systems more sophisticated than those of bumble bee colonies, I figured that honey bees have a very complex story of colony organization for energy economics. Of course, much was known already about the social organization of honey bee colonies, especially the famous dance language by which foragers recruit hive mates to rich food sources. This communication system had been deciphered in the 1940s by the Nobel laureate Karl von Frisch. I admired greatly Karl von Frisch's rigorous field experiments and his discoveries, but I also suspected that the bees possess additional wonderful "secrets" about how they cooperate to gather their colony's food. This suspicion proved correct, and even more so than I imagined at the time.

The most important thing that I learned during my Ph.D. and postdoctoral studies was that observing a biological phenomenon broadly is an invaluable first step toward understanding it. You get a "big picture" view of your subject and often you spy curious things that can lead you into new scientific terrain. I knew, therefore, that the best way to begin investigating how a colony is organized to gather its food was to describe the foraging behavior of a full-size colony living in nature. So Kirk and I teamed up to determine the spatiotemporal patterns of the foraging activities of a colony living in a forest. This is the natural environment for the honey bees found in North America, for these bees are native to Europe, whose lowlands were covered with forests for most of the last 12,000 years (since the last Ice Age).

To monitor a colony's foraging operation, we installed a full-size (ca. 20,000-bee) colony in a large, glass-walled observation hive, shown in Figure 12.1. Next, we loaded our large hive into the bed of a pickup truck and moved it to a hut we had set up in the center of a large forest owned by Cornell University. This forest, which is called the Arnot Teaching and Research Forest, lies about 15 miles (24 kilometers) southwest of Ithaca and covers about 7 square miles (ca. 20 square kilometers). It lies within the vast expanse of hilly woodlands (the Appalachian Highlands) that cover much of southern New York State. Figure 12.2 provides a glimpse of the Arnot Forest and the forest-covered hills that surround it for many miles.

We left the colony alone for about a week, to give it time to become familiar with its new location. Then we monitored the waggle dances of

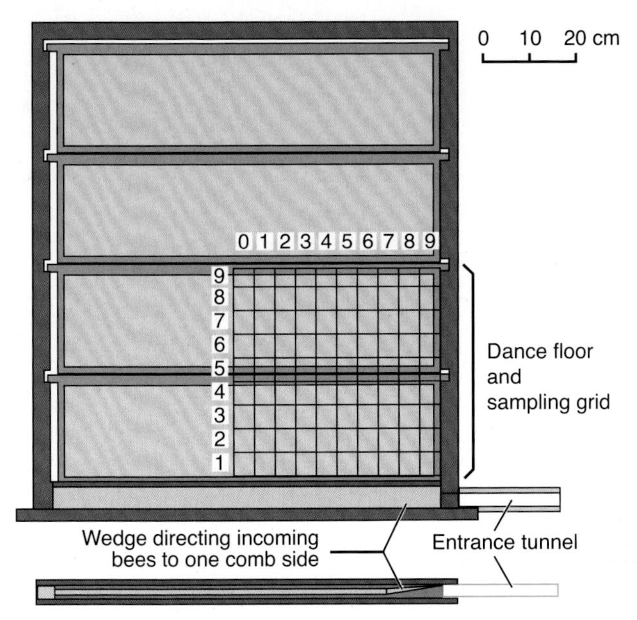

FIG. 12.1. The large observation hive that was used for sampling and reading the dances of forager bees in a full-size colony, to determine where this colony's foragers were gathering their food. A "traffic deflector" in the hive forced the foragers to enter and exit the hive from just one side of its four large combs. Therefore, this colony's foragers performed nearly all their waggle dances within the area of the sampling grid that is shown. Note: 20 centimeters = ca. 8 inches.

FIG. 12.2. A view to the southeast from a lookout point on the Irish Hill Road in the Arnot Forest. Photo was taken in early October, when the fall colors of the leaves were at their peak.

its foragers continually from 8:00 a.m. to 5:00 p.m. each day for four sequential days in June 1979. We could tell whether a dancing bee was advertising a patch of flowers that provided pollen, not just nectar, by noting whether there were, or were not, pollen loads on her hind legs. This pilot study helped us to develop our methods of data collection, and it gave us preliminary results that looked very promising. It was followed by a full-scale study in 1980, when Kirk and several helpers recorded the bees' dances each day for four 9-day periods during June, July, and August. The data collected in 1980 revealed the directions and distances of the flower patches that our colony's foragers were advertising to their nestmates on each day. By plotting the locations of these flower patches on circular maps, with the hive at the center, we were able to track, day by day, the foragers' most important recruitment targets as if they were objects on a radar screen. A sample of our results is shown in Figure 12.3.

The data collection and analysis were done entirely by hand—we had no video equipment and no personal computers—so the work was tedious. But it was also extremely rewarding, because it revealed that a honey bee colony functions like a gigantic amoeba that is fixed in one spot (its home) but is able to extend "pseudopods" (groups of foragers) out across the landscape to patches of flowers bearing nectar or pollen, or both. Note: if you are a beekeeper, then this finding has great relevance. It shows you that the foragers in your colonies can operate over a vast area around their hives, so even if the landscape right around your apiary is not blessed with plentiful flowers, your colonies can still thrive given the ability of their foragers to forage widely.

The foragers in our study colony made their forays over an area greater than 40 square miles (ca.100 square kilometers). The average distance to their flower patches was 1.2 miles (2 kilometers), but some were more than 6 miles (10 kilometers) away. (Note: 6 miles for a honey bee is the equivalent of about 600 miles for a human being, given the roughly 100-fold difference in body length between honey bee and human being.) We also learned that the bees' waggle dance communication system enabled

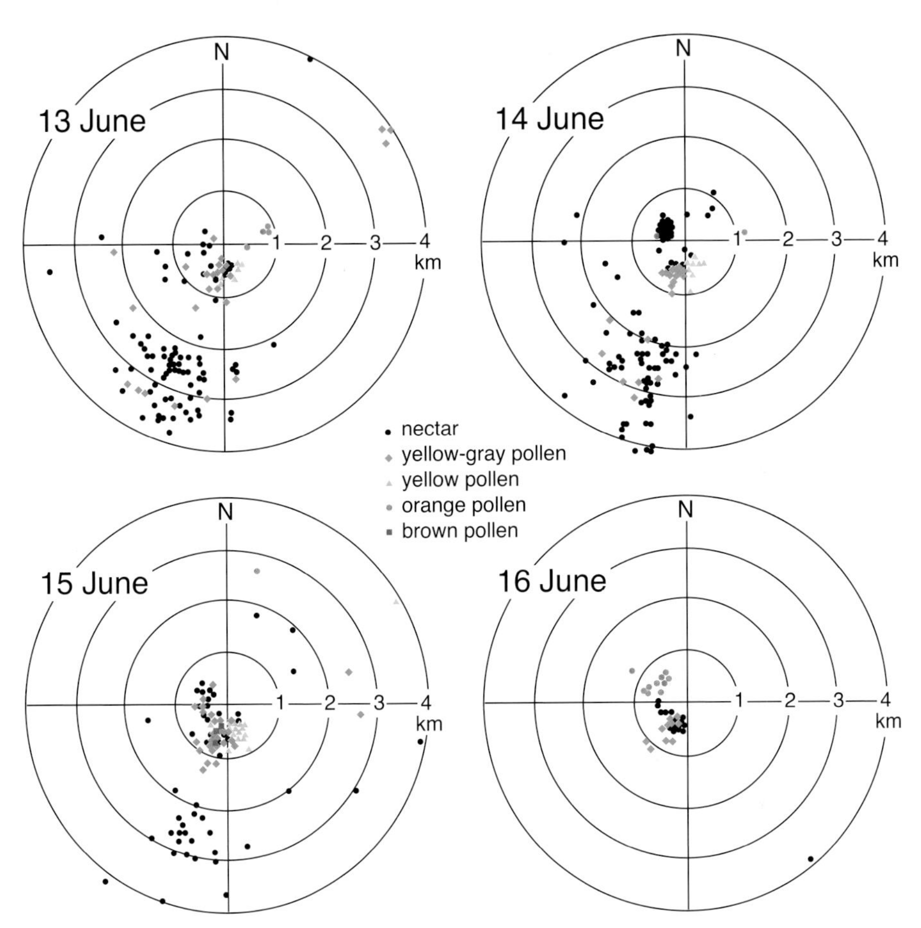

FIG. 12.3. Four maps of a colony's foraging sites on four consecutive days in 1980, as inferred from reading the waggle dances performed by the colony's foragers. Each dot shows the location indicated by one bee's dance. Black dots denote sites that provided nectar; colored dots denote sites that provided pollen of the color shown. Only a small fraction (2%) of the foragers' dances indicated sites beyond 2.4 miles (4 kilometers), and most of these sites are not shown.

our study colony to adjust daily the foci of its foragers' work, presumably to track the richest flower patches across the ever-changing landscape. For example, we see in Figure 12.3 that on June 13, the colony's nectar foragers danced strongly and thus recruited vigorously to a large nectar source—probably tulip poplar (*Liriodendron tulipifera*) trees in bloom—in a valley and ridge to the SSW and more than 1.2 miles (2 kilometers) from the hive. We also see in Figure 12.3 that on the next day, June 14, the colony's nectar foragers began recruiting strongly to a new nectar source to the NW and only about 0.3 miles (0.5 kilometers) away. Also striking is how on June 16, which was a cool day with rain showers, all of the colony's recruitment targets were less than 0.6 miles (1 kilometer) from the hive. Moreover, on this day, a source of orange pollen to the NW suddenly became a major recruitment target. Finally, over the period of June 17–19 (not shown in Figure 12.3), the foci of the bees' dances, and therefore of the colony's foraging, shifted from mostly in the west to mostly in the east. Thus we saw that every day there were big changes in the sprawling "flower market" around our study colony's home, and that its foragers were, evidently, tracking these changes.

Seeing these ever-changing maps of the targets of recruitment by the foragers in our study colony led me to start thinking of a honey bee colony not just as a *resource center*—a place where thousands of bees build their combs, rear their brood, and store their food—but also as an *information center*. In other words, I began to appreciate more fully than before that when a forager arrives home, she carries not only food in her pollen baskets and in her honey stomach, but also information in her brain. This is information about the direction, distance, and desirability of the flowers she has just visited. She can then share this information with her colony mates. To the best of my knowledge, only human beings and honey bees have the ability to steer fellow members of their group to food bonanzas by providing *abstract information* about their locations. In every other species of group-living animals, individuals must lead their groupmates to a rich find either directly or indirectly—that is, either by taking them to it or by marking the way with "signposts" such as the chemical trails laid by some ants. Once I started to think about a honey bee colony as an information center, I began to see many new mysteries about how a colony works.

The mystery that I decided to tackle first was this: How do the foragers in a colony solve the never-ending problem of deploying themselves wisely among the kaleidoscope of flower patches around their home?

I knew there is no "supervisor bee" that has an overview of a colony's foraging opportunities and directs the colony's foragers to their work sites each day. So I knew that the foragers must solve this problem among themselves, hence in a decentralized way. It was also clear that the waggle dance communication system makes it possible for a honey bee colony to adjust daily the deployment of its foragers and thereby track the richest flower patches across the ever-changing floral landscape. But how, exactly, does this tracking process work?

A colony's foragers can fly more than four miles (6.5 kilometers) to find food. This meant that to conduct experiments for analyzing how a colony deploys its foragers among food sources, I would need a study site surrounded by at least 50 square miles (130 square kilometers) of heavily forested countryside. I figured that a colony taken to such a site would experience times when its foragers would find few flowers and therefore would be eager to visit sugar-water feeders to keep their colony well supplied with food. This would make it possible to perform experiments to investigate how a colony's foragers adaptively allocate themselves among the "nectar" (sugar water) sources that I would provide.

Luckily, I found the vast flower desert that I needed. It is the expanse of forests, wetlands, and lake waters around the Cranberry Lake Biological Station (CLBS), which sits within the 9,375-square-mile (24,280-square-kilometer) Adirondack Park in northern New York State. I learned about the CLBS in the summer of 1980, during a chat with a beekeeper friend at Cornell, Dr. Jon Glase, who ran the large Investigative Biology Laboratory course that is mandatory for Cornell students studying biological science. Kirk Visscher and I were telling Jon about the study we were conducting in the Arnot Forest when Jon told us that he, too, had once taken a colony of honey bees in an observation hive to a forested setting, the CLBS, for teaching a course on animal behavior. He shared with us that when he had set out sugar-water feeders so his students could do experiments on honey bee color vision, the foragers from his colony had mobbed his feeders. Wow! The CLBS sounded most promising as a future site for experimental studies of the social organization of foraging by honey bees.

I began working at this lovely biological station in 1985 and was delighted to find that it is surrounded for 10 miles (16 kilometers) almost entirely by forests, beaver meadows, bogs, and the open waters of Cranberry Lake. There are no fields, few roadsides, and almost no gardens in the area, for nearly all the land around this immense lake—the hiking trail around it is 50 miles (80 kilometers) long—is owned by New York State. Just a few dozen summer houses ("camps") with flower gardens dot this lake's forested shores. The bee forage is so scanty that no wild colonies of honey bees live up here, and even colonies of bumble bees are scarce. The CLBS was (and still is!) a wonderful find. The rarity of natural nectar sources here means that any colony that I bring here must gather most of its "nectar" from my sugar water feeders, and this means that I have control of the colony's "nectar" sources. Now I had a field site where I could perform experiments to investigate how a colony's foragers keep themselves wisely allocated among the available nectar sources, despite never-ending changes in the floral landscape.

Among the early studies conducted at the CLBS was one that I hoped would give us, at last, a crystal clear picture of a colony's ability to allocate its foragers wisely across a dynamic array of foraging opportunities. In the study conducted in the Arnot Forest with Kirk, we had monitored changes in the number of *waggle dances that advertised* different foraging sites; now, I wanted to monitor changes in the number of *forager bees that exploited* different foraging sites. This was a tricky experiment, for it required working with a colony of 4,000 worker bees, each of whom was labeled for individual identification. I prepared such a colony in the summer of 1989, with the help of two summer assistants—Scott Kelley and Samantha Sonnak, both undergraduate students at Cornell. We chilled worker bees in groups of about fifty; then we glued a plastic tag (each with one of 500 number/color combinations) to each bee's thorax and applied a dot of paint (one of 8 colors) to each bee's abdomen, as shown in Figure 12.4. Once we had labeled each batch of 50 or

Fig. 12.4. Worker bees labeled for individual identification, with plastic tags and paint dots.

so chilled workers, we put them in a cage at room temperature with their queen and a feeder bottle with sugar water. The chilled workers would warm up, eat some food, cluster around the queen, and feel "queenright." At the end of two 12-hour workdays, our cage held 4,000 individually identifiable worker bees, plus their queen. We then installed all these bees in a two-frame observation hive that we had stocked with combs that contained young brood and considerable pollen, but little honey. We gave our bees only a meager honey reserve, so they would be eager to exploit our sugar-water feeders.

After labeling the bees in my laboratory at Cornell, we screened the entrance of the observation hive that housed the colony of labeled bees, and then we moved it 150 miles (240 kilometers) north to the Cranberry Lake Biological Station. Next, we trained 10 bees to collect sugar water from each of two feeders located in opposite directions (north and south) from the colony's hive (Fig. 12.5). Each feeder was set up in a sunny clearing a quarter of a mile (400 meters) from the hive. During the training

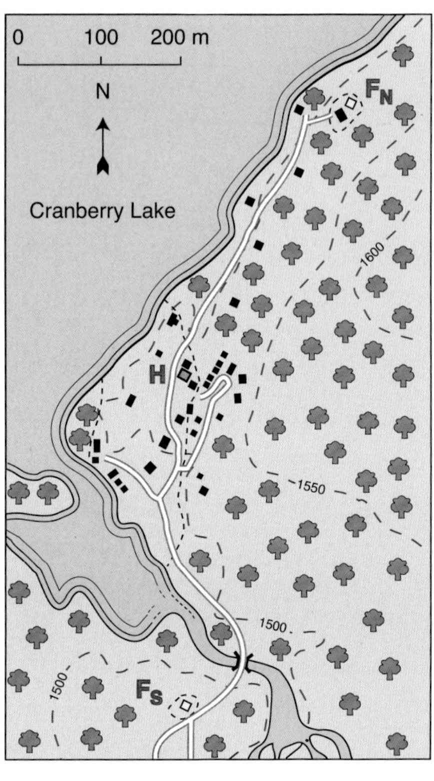

FIG. 12.5. *Right.* Layout of the experiment at the Cranberry Lake Biological Station in June 1989: H, observation hive; F_N and F_S, feeding stations a quarter mile (400 meters) north and south of the hive. Contour lines show feet above sea level. Tree symbols indicate areas with forest cover. Both feeders were set up in sunny clearings about 100 feet (30 meters) across. *Left.* The two-frame observation hive, mounted in the portable hut, that housed the colony of 4,000 individually identifiable bees. Bees entered this hive through a tunnel leading from the Plexiglas window in the hut's far wall. The hive was suspended from a metal bar overhead.

period, both feeders were filled with a rather dilute (30 percent) sucrose solution. This was rewarding enough to motivate the bees visiting each feeder to continue doing so, but it was not rich enough to stimulate these bees to perform waggle dances to recruit hive mates to our feeders. The critical observations began at 7:30 a.m. on the sunny morning of 19 June 1989. This was the first fair-weather day after a 10-day stretch of cool, rainy weather, during which the colorful bees in our little colony had not left their cozy hive.

After loading the north and south feeders with 30% and 65% sucrose solutions, respectively, we began recording the color-number ID code of every bee that visited our feeders during every half-hour block of time between 7:30 a.m. and noon. By midday, the colony's foragers had produced a strong pattern of differential exploitation of the two foraging sites: 91 bees were bringing home food from the richer feeder to the south, but only 12 bees were doing so from the poorer feeder to the north (Fig. 12.6). We then swapped the locations of the richer and poorer feeders for the afternoon, and by four o'clock the colony had switched the focus of its foraging efforts from south to north. The colony's collective wisdom in choosing between the food sources was observed again the next day in a second trial of the experiment. Thus, we saw that when we gave our study colony a choice between two "nectar" sources that differed in profitability, the colony consistently focused its collection efforts on the richer, more profitable source. The net result was that the colony's foragers tracked the richest foraging site within our changing array.

Figure 12.6 shows that a colony of honey bees is able to focus its foraging on the best of the nectar (or sugar water) sources its foragers have found. However, it does not show what the individual foragers in this colony did inside their hive to produce this colony-level "wisdom" in distributing themselves between the two food sources. Subsequent observations and experiments revealed the actions that a colony's nectar foragers take so that they wisely exploit the nectar sources they have discovered.

First, a nectar forager who is already engaged in working a flower patch (i.e., who is an "employed forager") adjusts several variables of her behavior in accordance with the energetic profitability of the patch. The greater its profitability, then the faster she works; the larger the nectar load she collects; the less likely she is to stop visiting the patch; the more likely she is to advertise it with a waggle dance; and, if she performs a dance, the more persistently she dances. For example, at the start of the experiment depicted in Figure 12.6, 80 percent of the bees that returned from the southern feeder, filled with the 65% sucrose solution, performed a waggle dance, but only 10 percent of the bees from the northern feeder, filled with the 30% sucrose solution, did so. Also, the durations of the waggle dances performed by bees returning from the south and north feeders differed markedly: about 20 seconds

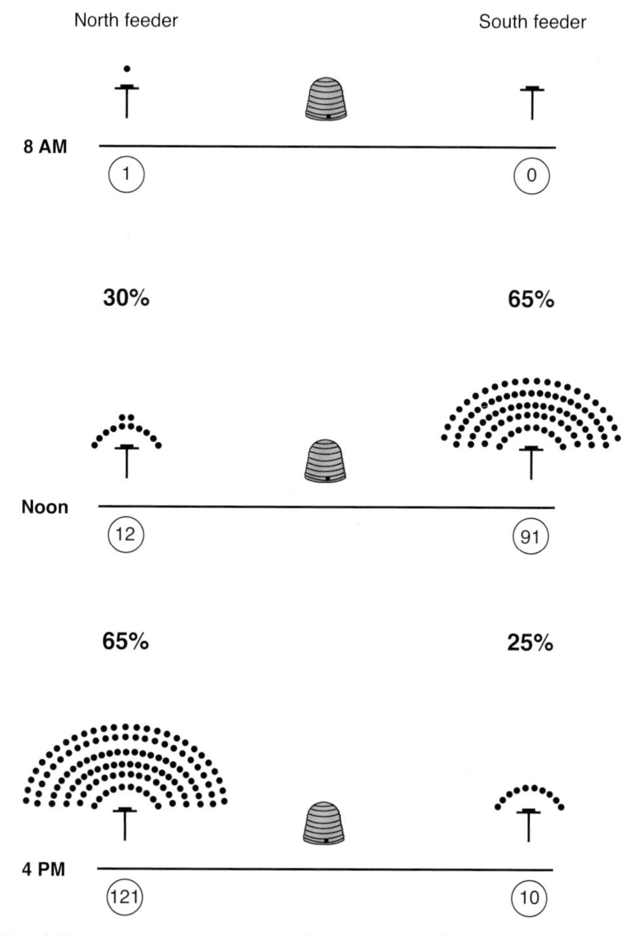

FIG. 12.6. The ability of a colony to choose between two foraging sites (sugar-water feeders) with either a 30% (or 25%) or a 65% sucrose solution. The number of dots above each feeder indicates the number of individuals that visited the feeder in the 30-minute period preceding the time shown on the left. The two feeders were located 0.25 miles (400 meters) from the hive and were identical except for the concentrations of their sugar solutions.

vs. 2 seconds, respectively. This shows us that the "dance floor" inside a colony's nest—i.e., the comb areas just inside a hive's entrance—functions like a help-wanted board where only highly rewarding employment opportunities are posted abundantly and persistently.

Second, a nectar forager who is unemployed and is looking for work will go to the dance floor and will follow a waggle dance *chosen at ran-*

dom to learn where she should go to find work. By choosing at random which dancer to follow, an unemployed forager ensures that her colony's forager force is deployed wisely among the sites being advertised. More bees are recruited to highly profitable sites than to less profitable sites, because the more numerous and longer-lasting dances for the richer sites are more likely to be encountered by unemployed foragers than the fewer and briefer dances for the poorer sites.

In later chapters, we will explore further the behavioral complexity of nectar foragers. We will see that besides the waggle dance that plays a key role in allocating these bees among work sites, the nectar foragers have several other signals—the tremble dance, the shaking signal, and the stop signal—that also play important roles in making a colony function as an information center.

Our studies of how the nectar foragers in a honey bee colony nimbly and wisely allocate themselves among flower patches, in face of never-ending changes in the locations of the richest ones, turned out to have immense practical value about 15 years later. What my co-workers and I had discovered inspired a team of engineers at the Georgia Institute of Technology to create the Honey Bee Algorithm (HBA). The HBA is widely used in cloud data centers (analogous to hives) to optimally allocate their server computers (analogous to foragers) among jobs (analogous to flower patches). This means that the Honey Bee Algorithm is integral to the multibillion-dollar industry of cloud computing, which underlies everything that you or I do on the internet. In 2016, I and four professors of engineering at Georgia Tech—John J. Bartholdi III, Sunil Nakrani, Craig Tovey, and John Vande Vate—were chosen by the American Association for the Advancement of Science to receive its Golden Goose Award for that year. This award recognizes esoteric research that proves extremely valuable (i.e., lays lots of "golden eggs"). What I like most about this award is that it highlights how our knowledge of honey bee behavior greatly enriches our lives, and does so in surprisingly diverse ways.

CHAPTER 13

Foragers as Sensors

In late summer, when I look across the hayfield that stretches southward from my barn, I can just barely see the bright-red fruit on the crabapple trees growing along the field's southern edge. These trees are 180 yards (165 meters) away, so I figure that this is the maximum distance at which I can spot this food, if I stand in one place. My range of detection is respectable, but it is dwarfed by the distance over which a honey bee colony, which always "stands" in one place, can spot its food. This is because my food-spotting organs (my eyes) are anchored in my body, whereas a colony's food-spotting organs (its foragers) can fly from their home and search widely. We know that foragers can zip off to flower patches six miles or more (10+ kilometers) from home. We also know that these bees can perform waggle dances to inform their hive mates of the whereabouts of flowers bearing plentiful nectar and pollen. This means that the foragers in a honey bee colony function as a powerful sensory system, one that supplies the colony with information about good food sources within a 100-square-mile (260-square-kilometer) area around its home. In this chapter, we will look at how a nectar forager functions as a nectar-source sensor for her colony, and so enables her colony to respond adaptively to the ever-changing floral landscape.

In June 1992, I headed off to the Cranberry Lake Biological Station (CLBS) in northern New York State. With me were three undergraduate students. Two were from Cornell University, Tim Judd and Barrett Klein,

FIG. 13.1. The southern end of Cranberry Lake, and the forest-covered landscape around it, as seen looking south from atop Bear Mountain. The yellow arrow marks the location of the Cranberry Lake Biological Station. It is surrounded by "Wild Forest and Wilderness" land within the 9,375-square-mile (24,000-square-kilometer) Adirondack Park. This is the largest park in the United States outside of Alaska.

and one was from the University of Zürich, Cornelia König. Our plan was to investigate how the foragers in a honey bee colony function as its sensors of high-quality nectar sources. We had loaded my pickup truck with three colonies of honey bees, a two-frame observation hive, the portable hut inside which I mount my observation hive, and all the rest of the paraphernalia (scientific and personal) that we would need to work at this remote field station for the next six weeks. We motored north for three hours to reach the marina of the CLBS. Here we transferred our gear to the station's ferryboat, the *Forester*, to make the four-mile (6.5-kilometer) trip across Cranberry Lake, the third largest lake within the vast Adirondack Park.

The CLBS sits in splendid isolation on Barber Point, surrounded by the lake's waters to the north, and by forests, bogs and ponds to the south, east, and west (Fig. 13.1). Its primary mission is to introduce undergraduate students to the organisms (primarily the plants, animals, and fungi) that live in the wilderness lands of the Adirondack Mountains, but it also supports field biologists with research projects that take advantage of its special setting. Because the CLBS has cabins and a dining hall, my team and I did not have to worry about our housing or our meals, so we were able to focus on our studies. After suppers and on rainy days, though, we had time for canoeing and hiking, as well as (of course) analyzing the data generated by our experiments. That the CLBS is "family friendly" made it doubly attractive to me as a study site. Every summer when my daughters, Saren and Maira, were young, they and Robin joined me there for several weeks of fun: swimming, canoeing, and having adventures with the children of the other folks working at this biologist's paradise in the North Woods.

A good way to start thinking about how the foragers in a honey bee colony function as its food-source sensors is to consider where in their nest (or hive) these bees perform their waggle dances to report the rich sources of nectar and pollen that they have encountered. They do so within a few small areas of comb just inside their nest's entrance. I call these places the "dance floors" in a colony's nest. Figure 13.2 shows the area of the dance floor in the two-frame observation hive—which has just one layer of comb—that I have used for many of my studies. By concentrating their dances in a few small areas near the entrance, the *employed* foragers make it easy for their colony's *unemployed* foragers to find dancing bees and get information about good places to work.

How is the information about different nectar sources arrayed on the dance floor? Figure 13.2 shows the locations of the waggle dances performed in my observation hive in the late morning on 29 June 1992. These dances advertised two widely separated sources of food. The nearby one was a sugar-water feeder. The distant one was a patch of raspberry (*Rubus* spp.) and fireweed (*Chamaenerion angustifolium*) plants that had sprung up where a logging operation had created sunny openings in the forest. The two recruitment targets were nearly the same direction from the hive (SSE and SE), but they were more than 6,000 meters (3.6 miles) apart. That some of the foragers in our study colony flew to flowers 6,500 meters (4 miles) away shows us that this colony's foragers had not found

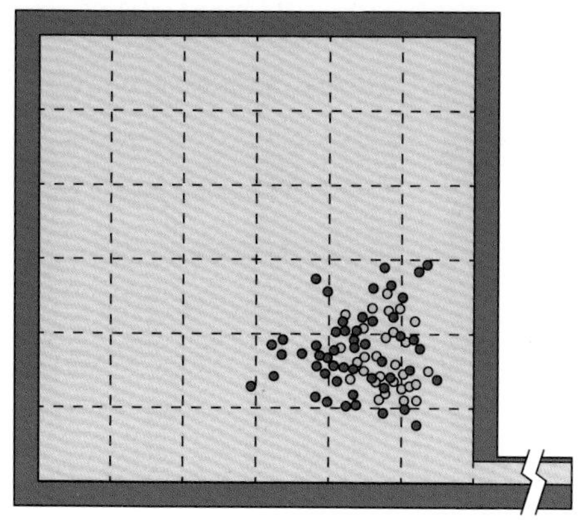

FIG. 13.2. Spatial distributions of the dances for two widely separated food sources. The two arrows indicate the orientation of the waggle-runs in the dances for the two sites. We see that the bees' dances for the two foraging sites were not spatially segregated. Dances were plotted during scan samples made at 2-minute intervals over a 20-minute period. A wedge (not shown) in the hive's entrance forced the bees to enter (and exit) the hive on just the side of the comb that is shown. 10 centimeters = 4 inches.

adequate food sources nearby. Figure 13.2 also shows us that the two groups of foragers, despite working at two widely separated foraging sites, performed their dances side by side, in a mosaic pattern, on the dance floor.

Why don't worker bees keep the information about distinct foraging sites separate when they perform their dances? I am pretty sure that they don't do so because interspersing the dances for different sites helps a colony achieve an optimal distribution of its foragers among the available foraging sites. We know that when a forager follows a dance to find a work site, she does not survey the dances being performed on the dance floor before leaving the hive to start her work. Instead, she follows just one dance, chosen at random from the jumble of dances present at the time. This random sampling helps ensure that a colony's foragers distribute themselves appropriately among multiple foraging sites, with more of them, but not all of them, working the larger and richer—and more strongly advertised—sites. Studies have shown that distributing foragers in this way produces an efficient collection of resources from the kaleidoscope of flower patches spread across the landscape.

It was clear from the work of Karl von Frisch that the waggle dance of a nectar forager contains information about the *direction* and *distance* to her foraging site. It was not clear, however, that a nectar forager's dance also contains information about the *desirability* of her work site. So we studied this in the summer of 1992. We now know that a forager's dance does indeed contain this third piece of information. How does this work? A nectar forager encodes information about the desirability of her work site by adjusting the duration of her dance. More specifically, she adjusts the number of waggle runs that she produces when she advertises her site. This is shown in Figure 13.3. It shows how the waggle dances performed by seven bees visiting a sugar-water feeder varied in duration from 1 to about 50 waggle runs/dance—in relation to the energetic profitability (joules of energy gained per joule of energy expended) of foraging at this feeder.

How did we obtain the data shown in Figure 13.3? On 29 June 1992, we video recorded the waggle dances that seven bees performed to report on a sugar-water feeding station that we had set up. It sat 1,150 feet (350 meters) north of our observation hive. We allowed only 12 bees from our study colony to forage at this feeder, and each one was labeled with paint marks for individual identification. (To prevent the feeder from becoming crowded, we captured the bees that were recruited to it, i.e., the bees without paint marks. These bees were released at the end of the day.) The experiment started when we filled the sugar-water feeder with a moderately concentrated (42% by weight) sucrose solution. We then video recorded the dance floor of the observation hive for 60 minutes to record the waggle dances of the bees that were visiting our feeder. Seven of the 12 bees performed waggle dances. During these 60 minutes, each bee made at least 10 round trips to the feeder; meanwhile, my helpers and I made the measurements we needed to know in order to calculate the energetic gain and the energetic cost of a trip to and from the feeder by each of the

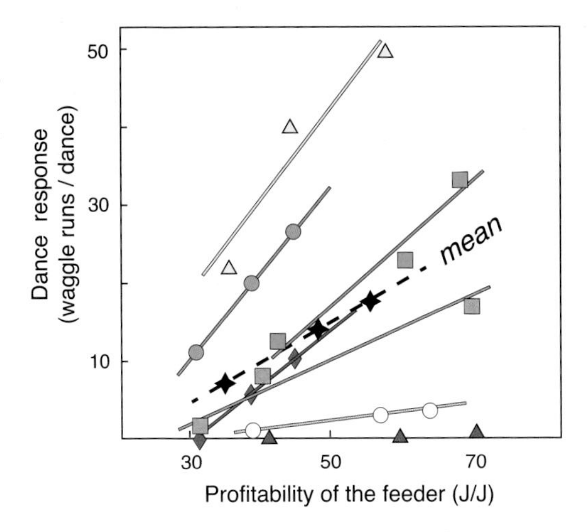

FIG. 13.3. How seven worker bees made reports on a feeder containing a sucrose solution whose concentration was set at three levels (42%, 56%, and 65% sugar, by weight), in three 60-minute periods, spread over an afternoon. By making various measurements of each bee's weight and behavior, I was able to calculate the energy efficiency (joules of energy gained per joule of energy spent, "J/J") that she experienced when she made a trip to the feeder. When each bee returned to the hive, a video camera recorded her dance response: the number of waggle runs she performed to advertise the feeder. Each bee was identified by paint marks on her thorax and abdomen. This graph shows that each bee linearly increased the duration of her dances as the profitability of the feeder was increased.

seven bees that advertised it by performing waggle dances. So, we measured the weights of the sugar-water loads of each bee (which determine how much energy she gained on each foraging trip) and the durations of each bee's feeder-bound and homebound flights (which determine the energy she expended on each foraging trip). Getting this information required weighing each bee when she arrived at the feeder and when she left the feeder, to see how much sugar syrup she collected. It also required recording the times at which each bee departed from the observation hive and arrived at our feeding station, and likewise when she departed the feeding station and arrived back at the observation hive. Then we did everything twice again, when the feeder was reloaded with a 56% and then a 65% sucrose solution. (*Technical note: because weighing the bees required handling them, and this altered the time course of their foraging and affected their dancing, we did all the weight measurements during a repetition of the experiment on the following day, when we did not collect data on their dancing.)*

This was a technically challenging experiment, but it was worth overcoming its difficulties because it revealed that each bee linearly increased the mean number of waggle runs per dance as the energetic profitability of the sugar-water feeder increased. This linearity makes sense for these bees. They had a broad, 50-fold range of possible dance responses to the feeder (from 1 to about 50 waggle runs per dance), but they experienced a narrow, 2.3-fold range of energetic profitabilities (from 30 to about 70 joules gained per joule expended) in working the feeder. In other words, the dance response range of 1 to 50 waggle runs per dance gave the bees plenty of range to indicate clearly when the profitability of foraging at the feeder was only so-so, or was pretty good, or was first-rate.

This experiment also revealed that there was great variation among individual bees in the thresholds of nectar-source profitability that elicited waggle dancing. For example, for the bee GG (whose dance responses are represented by the green line in Figure 13.3) this threshold was low, about 20 joules of energy gained per joule of energy expended, but for the bee WW (represented by the gray line) it was much higher, about 40 joules gained per joule expended. I believe that this variation among individuals is beneficial to a colony, because it means that a colony has a broad "dynamic range" in its responses to food sources. A low-profitability nectar source will be advertised by just a few dancing bees, while a high-profitability one will be advertised by many dancing bees. (*Technical note: To determine the threshold level of profitability for dancing by any of the bees represented in Figure 13.3, you extend her dance-response line down to the 0 level of dance response. The point at which the bee's dance-response line hits the horizontal axis indicates the profitability level that is her threshold for dancing.*)

The work that Tim, Barrett, Cornelia, and I did at the CLBS in 1992 was valuable in a second way, too, for it caused me to think more deeply about the results of a study that I had conducted there two summers back. This study was a collaboration with William F. Towne, a professor in the biology department at Kutztown University in Pennsylvania. Will, too, focused his research on the behavior of worker honey bees, especially on how young bees carefully memorize the sun's compass direction—relative to landmarks around their home—for each time of day. Having this knowledge enables honey bees to share information about food-source locations using the waggle dance *even on days when the sky is so cloudy that these bees cannot see the sun or patches of blue sky.*

Will had joined me and my research team at the CLBS in July 1990. The team that summer included a future graduate student, James C. Nieh, and two Cornell undergraduate students, Kim Bostwick and Stephen Bryant. Our goal was to determine how a forager that needs to find a work site samples the information provided by the bees performing waggle dances in her colony's nest. We wondered, does an unemployed forager attend dances advertising various flower patches, compare these dances, and then respond preferentially to the strongest one? At the time, some biologists thought that a forager that is trying to find a work site conducts "comparison shopping" on the dance floor. It looked to me, though, that when a forager needs to find a work site, she attends just one dance, and that it might be a dance chosen at random.

To clarify this matter, we brought to the CLBS a colony living in an observation hive, and then we trained two separate groups of 30 bees to collect sugar water from two feeders that were positioned in opposite directions (north and south) from the hive. The layout for this experiment was the same as what is shown in Figure 12.5. Figure 13.4 shows what our two feeders looked like and how they worked. In each trial of the experiment, one feeder offered richer food than the other. We predicted that if an unemployed forager follows waggle dances advertising various flower patches, and if she can tell which dance represents the richest flower patch and then she heeds only this dance, then we should see two things: (1) an unemployed forager will follow dances performed by foragers from both of our feeders before she leaves the hive to start foraging;

FIG. 13.4. *Left:* Sugar-water feeder from which forager bees are collecting a sucrose solution. *Right:* Sectional view of the sugar-water feeder. The sucrose solution is enclosed in a glass jar so that its concentration remains constant. This glass jar sits on a grooved plate which gives the bees access to the sugar solution. The feeder is marked with a scent (anise) that evaporates from the reservoir of anise extract and escapes through the wire mesh beneath the grooved plate.

and (2) the dances for the richer feeder will be more effective, per waggle run, than the dances for the poorer feeder.

We found that neither prediction was correct. Each unemployed forager followed *just one bee's dance* before she flew from the hive. Also, the fraction of the recruits arriving at the richer feeder matched the fraction of the dance circuits produced in the hive to advertise this feeder. For example, in a trial when the north feeder was filled with a low concentration sucrose solution and the south feeder contained one of medium concentration, we found that the percentages of the recruits to the north and south feeders (12 percent and 88 percent) essentially matched the percentages of waggle-dance circuits that the bees produced for these two feeders (10 percent and 90 percent). These results, and others like them, told us that unemployed foragers do not conduct "comparison shopping" of the dances performed in their home. I think that this conclusion makes lots of sense. Almost always, a colony's success in foraging will be greater if its foragers distribute themselves among the various flower patches being advertised in the hive than if they crowd themselves on just one patch. Any given patch might be a dandy pollen (or nectar) source for a modest number of foragers, but not for a multitude of these bees. Apple orchards, blueberry barrens, and forage-crop (e.g., clover) fields are examples of places that are fine targets for mass recruitment, but they are all artificial. I believe that the social organization of foraging by honey bee colonies is what it is because for most of their history honey bees have lived in landscapes where first-rate food sources were fairly small, widely scattered, and highly ephemeral.

The forest-covered hills south of Ithaca, New York, are an example of this kind of landscape. Most of the forest vegetation that covers these hills does not provide forage for honey bees, but in some ravines and other places where the soil is rich, there are tall linden trees (*Tilia americana*) and stately tulip poplar trees (*Liriodendron tulipifera*). When these trees come in bloom, their white and yellow flowers can produce such sweet aromas and plentiful nectar that the bees seem to forsake all others. We saw a sign of this in the previous chapter, when I described how Kirk Visscher and I monitored where the foragers from our study colony in the Arnot Forest were working. On 13–15 June 1980, many of the dances in this colony were performed by nectar foragers working one distant location southwest of the hive. I discovered some years later, while hunting for wild colonies of honey bees in this forest, that this location is a shady woodland cove that is home to many handsome tulip poplar trees.

In 1992, I was thinking about how a colony's foragers endow it with a sensory system that is "designed" to help their colony detect and respond to desirable food sources. I realized then that I should review the data that had been collected in the summer of 1990 at the CLBS during our investigation of how unemployed foragers sample the waggle dances performed in their colony's nest. Might these data reveal something about how a nectar forager adjusts the strength of her waggle dancing for a nectar source in relation to the general availability of nectar? Many researchers, including Karl von Frisch, have reported that at times of the year when nectar is sparse, a nectar forager will dance vigorously to advertise a sugar-water feeder that offers only a dilute sucrose solution. It seemed, therefore, that nectar foragers follow the rule, "Beggars cannot be choosers." I wondered if the data collected in 1990 might show clearly and precisely that a nectar forager does indeed adjust her choosiness (i.e., her "stimulus threshold") for reporting a nectar source with a waggle dance, and that she does so in relation to nectar availability.

Back in 1990, we had established two feeding stations, one north and one south of the observation hive (see Fig. 12.5). The two feeders contained sucrose solutions with different concentrations, so they elicited waggle dances with different strengths (i.e., different durations). Every day, for 12 days (from 11 to 22 July, 1990), we video recorded the bees' dances for both feeders. We did so to determine how much dancing was done for each feeder, as described above. What I realized in 1992 was that the data extracted from these video recordings would tell me the mean dance duration for each feeder on each day. Moreover, I recalled that, back in 1990, I had measured each day, just before nightfall, the weight gain (or loss) of a full-size colony that I kept on scales at the CLBS that summer. These weight changes indicated the availability of nectar across the landscape around Cranberry Lake on each day. When we began our studies in early July 1990, there was a surprisingly strong nectar flow from the raspberry plants about 5 miles (8 kilometers) to the east of the CLBS, where a logging operation had created a clearing in the forest. I knew this from watching the waggle dances being performed in the hive; many of the dancing bees advertised this site and carried loads of raspberry-plant pollen. Thus, the scale-hive colony had gained weight at a rate of about 3 pounds (1.5 kilograms) each day at the start of our experiment. But by the end of our experiment, in late July 1990, the raspberry bloom was over, and our scale-hive colony had lost weight at a rate of about 1 pound (0.5 kilograms) each day.

Figure 13.5 shows the graph that I made based on my reanalysis of the 1990 data. It shows that when nectar became scarce in late July, the nectar foragers greatly lowered the threshold of nectar-source profitability that elicited waggle dancing. For example, on 11 July, when nectar was plentiful (and the scale-hive colony gained weight), a concentrated,

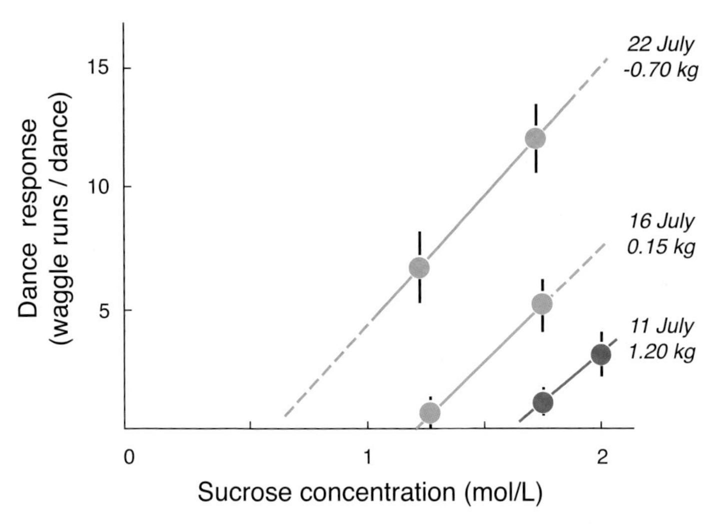

FIG. 13.5. How strongly the bees danced for the sugar water in the feeder depended on the availability of nectar from various flowering plants. Nectar availability was measured by noting the daily weight change of a colony living in a hive set on scales. The weight *gain* of 1.2 kilograms (2.6 pounds) on 11 July shows that nectar was moderately abundant that day, while the weight *loss* of 0.7 kilograms (1.5 pounds) on 22 July shows that nectar was scarce on this day. This graph shows that when nectar was scarce on 22 July, the bees danced vigorously even for sugar water with a low sucrose concentration.

2 mol/L (56%) sucrose solution elicited only a feeble dance response: just 2 waggle runs per dance, on average. But on 22 July, when nectar was scarce (and the scale-hive colony lost weight), a less-concentrated, 1.75 mol/L (42%) sucrose solution elicited a vigorous dance response: 12 waggle runs per dance, on average. These results show that when the nectar sources of a honey bee colony dwindle, its nectar foragers lower their dance-response threshold. In other words, they become less finicky about which flower patches they will advertise with their dances. This helps the colony to keep bringing in nectar, even though the nectar is not as rich as before. In other words, the nectar forager bees in a honey bee colony do indeed follow the rule "Beggars cannot be choosers."

In my studies of the bees, I have found that every successful investigation answers some questions . . . and reveals still more questions. For example, the findings depicted in Figure 13.5 solved one mystery *and* they raised another: How does a nectar forager know whether her colony is experiencing a nectar "flow" or just a nectar "trickle"? If you are a beekeeper, and you have a colony living in a hive set on scales, then you can assess this colony's foraging success each day—by noting in the evening how the hive's weight has changed since the previous evening. A nectar forager, however, can sense her colony's success in nectar collection on a much finer time scale: *each time she returns to the hive*. How she does this is the subject of our next chapter.

CHAPTER 14

Nectar Flow On?

When I was a teenager in the mid 1960s, and not yet a beekeeper, I discovered two bee hives near an old farmhouse down a dirt road, up on Snyder Hill. These hives sat in the lee of a tree line, about 100 yards (90 meters) from the house. A few days later, I returned to the farmhouse, knocked on the side door, explained who I was—a kid from down in Ellis Hollow who was curious about honey bees—and asked for permission to sit beside the hives and watch the bees. I must have seemed harmless, for permission was granted. This enabled me to observe closely how worker honey bees behave at their hive's entrance, and doing so fed my curiosity about these small but complex creatures. For instance, it was fascinating to see how they landed smoothly and then scurried into their hive, often bearing colorful loads of pollen on their hind legs. Likewise, it was intriguing to see how, on hot afternoons, a dozen or more worker bees stood in a neat line at their hive's entrance, all facing inward and fanning their wings. It was clear that these insects had sensed that their home was in danger of overheating. I remember especially well the warm September evening when I first smelled the peculiar aroma of goldenrod nectar (which I likened to the smell of sweaty socks) wafting from these hives. That the bees persisted in ventilating their home in the evening, when most of the foragers had stopped foraging, told me that on a day when a colony has made a large haul of fresh nectar, some of it members will work into the wee hours to "ripen" it into honey.

These experiences showed me that honey bee colonies do not build up their honey stores steadily. Instead, there are nectar "flows" when the plants that produce plentiful nectar come into flower, and there are times of nectar dearth. Where I live, the principal nectar sources between April and August are dandelions, white clover, blackberries, Japanese knotweed, and various sorts of shrubs and trees, including staghorn sumac, black locust, tulip poplar, and linden. In September and October, the main nectar sources are the stands of goldenrod and purple aster that shine brightly in waysides and abandoned fields. During these times of bee-forage plenty, I enjoy watching returning foragers plunk down at their hive's entrance and then dash inside with their abdomens almost dragging because their "honey stomachs" (crops) are stuffed with nectar. I hope that you, too, are able to enjoy this cheery sight.

In 1986, I decided to get a clear picture of the seasonal pattern of nectar abundance and nectar scarcity in the location where I had done several studies with honey bees in the early 1980s: the Yale-Myers Forest in Union, Connecticut. So, I set up two small lean-to shelters near the Forest's headquarters, and in each one I mounted a precise balance-beam scale on which I set a hive that housed a strong colony of bees. Also, I arranged with the logger and sawmill owner who lived just down the road, Mr. Travis Stevens, to visit these two hives each evening just before dark (when most of the bees were at home) and get a reading of each hive's weight. Figure 14.1 shows the daily changes in the weight of one of these hives—the weight of the hive per se, plus the weights of the adult bees, the brood, and the food stores inside the hive—across three months that summer. We see that this colony experienced a string of booms and busts in the supply of nectar.

Beekeepers are not the only individuals who pay close attention to a colony's success in gathering nectar. The nectar foragers in a honey bee colony do so, too. Indeed, we know that each time a nectar forager comes home, she gets an indication of her colony's foraging situation, and that she uses this information in deciding how strongly, if at all, she should advertise her nectar source by performing a waggle dance. The greater a

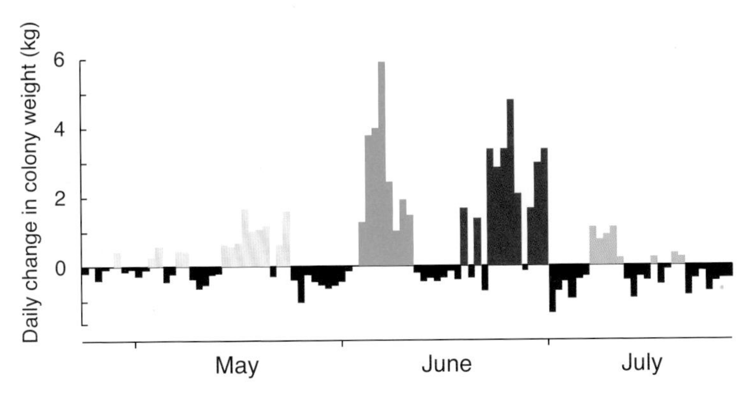

FIG. 14.1. Day-to-day changes in a colony's success in nectar collection. Between late April and late July, the colony experienced four nectar flows, that is, periods of intense collection of nectar. The principal nectar sources for these four flows were dandelions (May), black locust trees, (early June), sumac shrubs (late June), and basswood trees (early July). Between these times of large weight gains from nectar collection, the colony lost weight, either because plentiful nectar sources did not exist or because poor weather hindered foraging. Note: 1 kilogram = 2.2 pounds.

colony's intake of nectar, the more selective the nectar foragers are about advertising their sources. We saw a clear indication of this in Chapter 13 (Fig. 13.5). To recap, on 11 July 1990, my study colony discovered a large stand of raspberry (*Rubus* sp.) plants in bloom about 5 miles (8 kilometers) from the Cranberry Lake Biological Station, and many of its foragers flew there and brought home much nectar. Meanwhile, the foragers from my study colony that were visiting my sugar-water feeding station, which sat only a quarter of a mile (400 meters) away, refused to advertise this "nectar" source with strong waggle dances. Even after we had filled the sugar-water feeder with a highly concentrated sucrose solution (2.00 mol/L = 56% sucrose by weight), the bees foraging there performed only a few, and rather half-hearted, waggle dances. But 11 days later, on 22 July, by which time the nectar flow from the raspberry plants had abated, the bees visiting the feeding station advertised it excitedly. Now they performed long-lasting waggle dances even though it was filled it with a sugar solution that was less rich than before (1.75 mol/L = 49% sucrose). That they danced vigorously on this day made sense to me. Now that nectar-bearing flowers had become scarce, the bees had lowered the threshold level of food-source quality that elicited waggle dances, for doing so helped them to keep food coming into their hive. I find it easy to sympathize with the bees about this change in their behavior. When I come home and find little food in the house, I am not finicky about what I will have for supper.

Let's now consider the mystery of how a nectar forager adjusts the threshold level of food-source quality that elicits waggle dancing. What tells a nectar forager to not perform waggle dances for a sugar-water feeder when nectar is plentiful, but then to dance wildly for the same feeder when nectar becomes scarce? Karl von Frisch was vexed by this phenomenon when, in order to conduct an experiment, he needed the worker bees that were collecting sugar water from a feeding station to perform waggle dances in his observation hive, but the bees refused to do so. He wrote about this problem in his magnum opus, *The Dance Language and Orientation of Bees*:

> Success [in establishing an artificial feeding place] is threatened from two directions: in the spring the natural honeyflow is so good that, even with concentrated sugar solution to which honey has been added, it is hard to get bees to come to the food dish. They prefer the fields with their blossoming flowers. Often, all we could do was to put off the beginning of experimentation for several weeks. But in late summer, after the natural honeyflow has ceased, the colonies are so eager for anything sweet that strangers from other hives may become a pest and, unless great care is taken, will have driven off the experimental bees before one notices.

If you are a beekeeper, then you know all about this second scenario. When nectar is scarce, you must take measures, such as making your hives' entrances smaller, to help your colonies keep out would-be honey robbers.

Karl von Frisch's most distinguished student, Martin Lindauer, was curious about what causes a worker bee to be less likely to perform waggle

FIG. 14.2. A nectar receiver (left) has inserted her tongue between the mouthparts of a nectar forager (right). The receiver bee is imbibing the nectar that the forager bee is unloading from her swollen honey stomach (crop).

dances during a nectar flow (also called a "honeyflow") than a during a nectar dearth. He probed this mystery by installing a small colony in an observation hive and then training several of its foragers to collect sugar water from a feeder near the hive. He applied paint marks to each of these foragers to make them individually identifiable. Then he watched his marked bees when they returned to the observation hive. He saw that when these forager bees came home on days when nectar-bearing flowers were abundant, they searched longer to find bees willing to receive their loads of sugar water (Fig. 14.2) compared to days when these flowers were scarce. Lindauer supposed that this was because the nectar-receiver bees had raised the threshold level of nectar sweetness that was acceptable to them on the days when nectar was plentiful. In other words, he attributed the increase in the search times of his marked foragers to a rise in the *quality* (not the *quantity*) of the nectar coming into his study colony during a nectar flow. However, Lindauer did not show that nectar quality (i.e., its sugar concentration) is higher when nectar is more plentiful. It seems that he just assumed this. All of what Lindauer reported was plausible to me; it could explain why one sees exuberant waggle dancing in times of nectar scarcity but not in times of nectar abundance. So it was what I believed for many years.

In 1984, however, I realized that Lindauer might have been mistaken in his explanation of why nectar foragers are more reluctant to perform vigorous waggle dances during nectar flows than during nectar dearths. I saw that this might be due to the higher nectar *quantity*, not a higher nectar *quality*, during a nectar flow. In other words, when a nectar forager experiences a long search time to find a willing nectar receiver, perhaps this tells her that "A nectar flow is on!" rather than "Sorry, your nectar is subpar." What led me to reconsider Lindauer's explanation was thinking about checkout lines in grocery stores. Like everyone who goes grocery shopping, I had noticed that when customer traffic surges, one's waiting time to reach a cashier increases. One afternoon, while standing in a longish line, I related my situation to that of a nectar forager experiencing a delay to start unloading her nectar during a nectar flow. I knew that nectar foragers don't wait in lines like people do; instead, they (the "customers") walk around on the comb areas just inside the hive entrance and search there for nectar receivers (the "cashiers"). But I also knew that the two situations are fundamentally similar. In the lingo of operation research engineers, both are forms of "queueing."

At this point, I was eager to see if worker bees become less apt to perform waggle dances during a nectar flow because there has been an increase in the quantity, not the quality, of the nectar being collected by their colony. The first trial of my experiment that addressed this question was performed the following summer, at the Cranberry Lake Biological Station, a place where nectar is usually scarce (see Chapter 12). On the morning of 18 June 1985, two Yale students (Andy Swartz and Roy Levien) and I trained two groups of foragers to visit two sugar water feeders, one north and one south of the observation hive that housed our study colony, as shown in Figure 14.3. Each feeder was 1300 feet (400 meters) from the hive. And every bee that visited our feeders was labeled with a paint mark for individual identification. This way, we knew precisely how many bees were working each feeder.

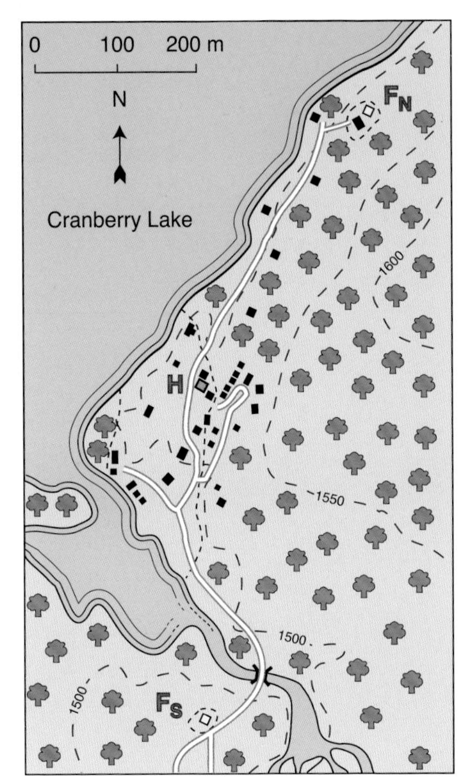

FIG. 14.3. Left: Map of the experimental layout at the Cranberry Lake Biological Station in June 1985. H, observation hive; F_N, and F_S, the feeding stations north and south of the hive. Right: Worker bee labeled for individual identification, collecting sugar syrup from one of the feeders. This bee's name was "Red/White."

From 1:00 p.m. to 3:00 p.m., we allowed 90 bees and 30 bees to forage at the north and south feeders, respectively, as shown in Figure 14.4. While my assistants tended the feeders, I sat beside the observation hive and watched the in-hive behavior of the bees that were making trips *to the south feeder*. There were two things that I recorded: (1) how long each south-feeder bee searched to find a nectar receiver when she returned to the hive, and (2) whether or not each of these south-feeder bees performed a waggle dance to advertise her feeder when she returned to the hive. (*Technical note: my assistants controlled how many bees were making collecting trips to each feeder. They did so by "jailing" any excess bees in Ziploc plastic bags. The bagged bees were released at the end of*

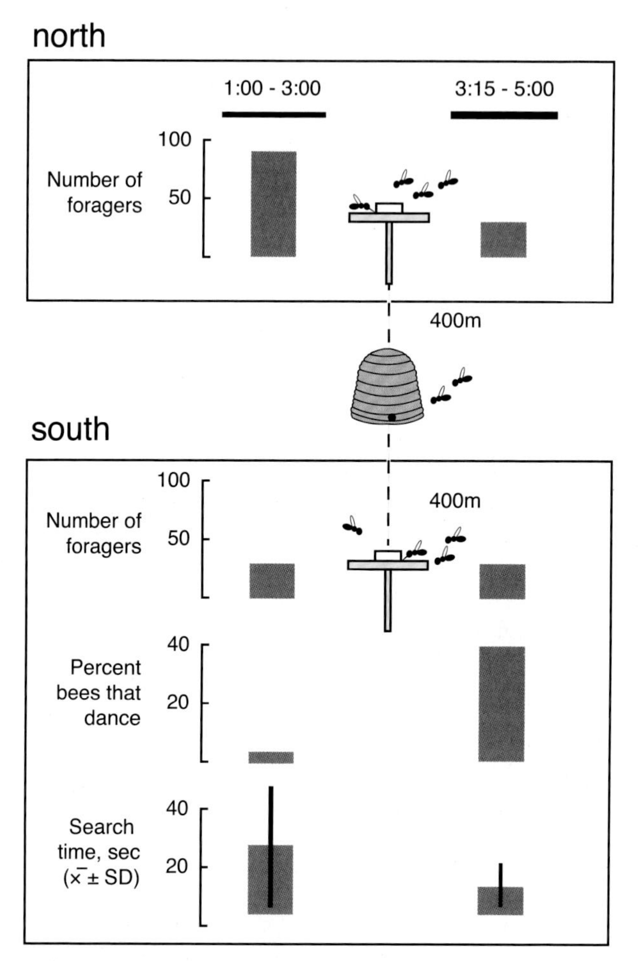

FIG. 14.4. The test of what causes nectar foragers to raise (or lower) the threshold of nectar-source profitability that stimulates them to perform waggle dances. Lowering the number of bees gathering sugar water from the *north* feeder (starting at 3:00 in the afternoon) triggered a drop in the in-hive search times of the bees working the *south* feeder. It also triggered a surge in the percentage of the south-feeder bees that performed waggle dances to advertise their feeder. This experiment showed that a colony's nectar foragers adjust the strength of their waggle dancing in relation to their colony's rate of nectar intake.

the day.) Then, at 3:00 p.m., Andy began to capture 60 of the bees that were visiting the *north* feeder. By 3:15 p.m., he had done so. This cut in half the colony's rate of "nectar" (sugar water) intake, and it did so without altering anything about the *south* feeder. Specifically, now there were only 60 bees, not 120 bees, bringing home sugar water from the two feeders, which were the primary sources of "nectar" for the study colony. Once Andy had finished capturing most of the north-feeder bees, I resumed recording the in-hive behaviors of the south-feeder bees. Almost immediately, the data I was collecting showed something that thrilled me: the percentage of the south-feeder bees that performed waggle dances had jumped from 3 percent to 40 percent, *even though nothing whatsoever had changed at their feeder!*

Two repetitions of this experiment, on 20 and 21 June 1985, yielded the same striking result: many more of the bees working the south feeder started performing waggle dances when the colony's influx of "nectar" from the north feeder was trimmed. It is important to note that in every trial there was no change in the *quality* of the food being collected; the only change was in the *quantity*. These results show that a colony's nectar foragers will shift their dance thresholds in response to a change in their colony's rate of nectar intake. When it goes up, so does the threshold of nectar-source profitability for waggle dancing. This experiment shows us that bees functioning as nectar foragers follow a built-in rule about producing waggle dances: when times are good, advertise only high-quality sources, but when times are not so good, advertise lower-quality sources, too. The functional significance of this behavioral rule is that it helps a colony to focus its nectar foragers on high-quality nectar sources when nectar is plentiful, but then to be less choosy about where they work when nectar is scarce.

The results shown in Figure 14.4 raise the question: How does a nectar forager know whether her colony's rate of nectar collection is high or low? The results regarding search times, shown at the bottom of Figure 14.4, show us one way that a nectar forager might do so: by noting how long she has to search in the hive to find a nectar receiver bee. When the total number of nectar foragers was 120 bees, between 1:00 and 3:00, the bees from the south feeder had to search in the hive for 28 seconds, on average. But between 3:15 and 5:00, when the total number of nectar foragers was just 60 bees, the bees from the south feeder experienced markedly shorter search times, just 14 seconds, on average. This is a phenomenon that is familiar to us all: how long we wait in a line to reach a cashier in a grocery store is a function of the level of customer traffic (unless the store manager decides to open up more checkout lines).

It was exciting to see that a nectar forager *might* get good information about her colony's success in nectar collection simply by noting how long she has to search in the hive to find a nectar receiver bee, i.e., the delay she experiences in starting to get unloaded. I asked myself, though, is the search time to find a nectar receiver the *only* cue that helps a nectar forager sense whether her colony is experiencing a low or high rate of nectar intake? Might she be paying attention also—or instead—to something else, such as the level of forager traffic at the hive entrance, or the level of the aroma of fresh nectar in the hive? In 1987, I went back to the Cranberry Lake Biological Station and performed some additional two-feeder experiments to address this question. I had moved from Yale to Cornell in June 1986, so in the summer of 1987 my research assistants were two Cornell students, Mary Eickwort and Oliver Habicht. The experiments we performed were similar to the one just described from the summer of 1985. But now we wanted to see if there are other variables of the unloading experiences of nectar foragers—besides how long they search to find their unloaders—that differ between times of high and low levels of nectar-forager traffic.

Our work in 1987 will be discussed more fully in the next chapter, so for now I will give just a preview of some of the things that we learned. *First*, this work confirmed that the *time* a nectar forager spends searching for a nectar receiver inside the hive is tightly correlated with her colony's level of nectar-forager traffic. *Second*, this work revealed that the *location* of the nectar unloading is strongly influenced by the traffic level of the nectar foragers. This is illustrated in Figure 14.5, which is based on observations made during an experiment in which we manipulated the traffic level of incoming nectar foragers. Initially, when their traffic level was high, each forager crawled at least 4 inches (10 centimeters) onto the combs inside the hive before she found a bee interested in receiving her nectar. But after we lowered the influx of the nectar foragers, each one made contact with a receiver bee either out in the entrance tunnel or

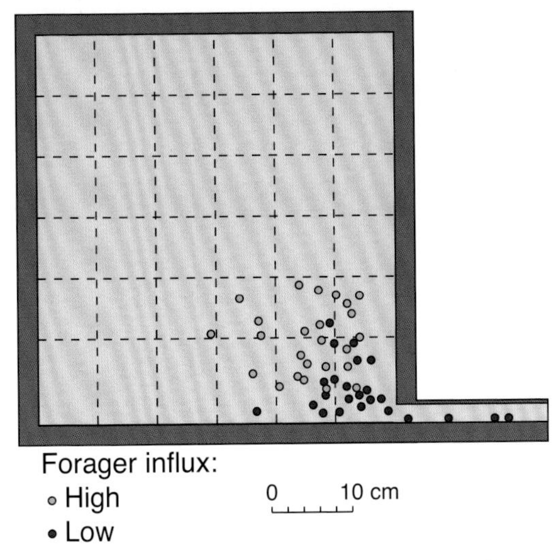

FIG. 14.5. Spatial distributions of the locations in the observation hive where nectar foragers passed off their nectar loads to nectar receivers. When the colony's rate of nectar collection was high, the foragers had to walk deep inside the hive to find bees willing to receive their nectar loads. But when the colony's rate of nectar collection was low, the foragers unloaded their nectar out in the entrance tunnel or just inside the hive. This figure shows that where a nectar forager gets unloaded can help her sense whether her colony's rate of nectar collection is high or low. Note: 10 centimeters = 4 inches.

just inside the hive. *Third*, this work showed that the maximum number of receiver bees that simultaneously unload a nectar forager is markedly different between periods of high and low rates of returning nectar foragers: 1.2 vs. 2.5 receiver bees per forager, respectively. Also, it looked like the nectar receivers were more "eager" to unload the nectar foragers when their traffic level was lower, but we did not have a reliable way to measure the eagerness of a nectar receiver. I hope that someday, someone equipped with sophisticated tools for video recording and video analysis will explore this topic further, and so will investigate more deeply the mystery of how the nectar foragers in a honey bee colony adjust their behavior to work productively during both times of abundance and times of scarcity.

I look back fondly at all the studies that I have done, but I have a special fondness for the ones described in this chapter. This is partly because these studies solved a mystery about honey bees that had stumped two of my "science heroes"—Karl von Frisch and Martin Lindauer—and partly because these studies have helped biologists understand something fundamental about animal communication. The key discovery (presented more fully in the next chapter) was that nectar foragers stay informed about their colony's rate of nectar intake without receiving a special signal that provides this information. Instead, the nectar foragers get this information by paying attention to a variable—the time spent searching for a nectar receiver—that is *an incidental by-product* of the interactions between the nectar collectors and the nectar receivers. This discovery bolstered an insight that another behavioral biologist, Professor James E. Lloyd, at the University of Florida, had recently reported back in the early 1980s. He pointed out that biologists who study animal social behavior need to distinguish between two markedly different ways that information flows between the members of a group: via *signals* and via *cues*.

A signal is an information-bearing structure or behavior that has been shaped by natural selection to convey information about a variable of interest. For example, in honey bees, when a worker releases a puff of alarm pheromone she indicates the presence of a predator. Also, when she performs a waggle dance she indicates the location of a rich food source. Both of these signals—pheromone puff and waggle dance—have been refined by natural selection over evolutionary time so that one bee provides other bees with information that is clear and accurate. Also, both pheromone puffs and waggle dances are produced intentionally by the sender.

A cue, however, is something that provides information clearly and accurately, *but does so incidentally rather than intentionally.* Two examples of cues for honey bees are a high level of gaseous carbon dioxide in the nest, which indicates a need for more ventilation, and a pile of unpacked (not yet tamped down) pollen loads sitting in a cell, which indicates a need for pollen packing in this cell. Neither cue has been shaped by natural selection to convey information; both are incidental by-products of colony functioning that happen to convey useful information.

The same is true for how long it takes a nectar forager to find a bee willing to receive her nectar. Her search time helps her decide whether she should or should not perform a waggle dance, and it does so incidentally.

Biologists who study animal social behavior used to focus almost exclusively on the conspicuous transfers of information among individuals via signals, to understand how a group's members interact. But now biologists know that we must also pay close attention to the more subtle forms of information transfer via cues. It is gratifying to know that my analysis of how the nectar foragers in a honey bee colony stay informed about their colony's rate of nectar intake—by sensing a cue rather than receiving a signal—promoted a general insight that has helped biologists understand better how animal societies work.

CHAPTER 15

Mystery of the Tremble Dance

For centuries, the meanings of the conspicuous shaking, waggling, trembling, buzzing, and piping behaviors of worker honey bees were deep mysteries. In 1788, for example, a German beekeeper named Ernst Spitzner wrote the following words after watching a bee perform a waggle dance: "Without warning, an individual bee will force its way suddenly in among 3 or 4 motionless ones, bend its head toward the surface, spread its wings . . . and execute a genuine round dance. . . . What this dance means I cannot yet comprehend." Spitzner's words show that he suspected that the dancing bee was sending a signal to her hive mates. His words also show that he was puzzled by the meaning of her behavior. This dance received no careful attention for another 130 years, that is, until the late 1910s. This is when a young Austrian zoologist, Karl von Frisch, began to investigate the dances of honey bees. Thanks to his wonderful studies, we know that when a worker bee performs a waggle dance she is sharing information about the location of something important: usually a lush patch of flowers bearing nectar or pollen, but sometimes a handy source of water or resin (for making propolis), and occasionally a potential homesite. In this chapter, we will look at another conspicuous, and long mysterious, kind of dance performed by worker honey bees: the tremble dance.

The tremble dance became a prime focus of the unknown in the biology of honey bees back in 1923. This is when Karl von Frisch published his first detailed report on the dances of honey bees, *Über die "Sprache" der Bienen (On the "Language" of Bees)*. In this 186-page account of his studies, he described not only the waggle dance, but also a second dance that he called (in German) *der Zittertanz*. In English, it is usually called the "tremble dance." I have translated von Frisch's description of this dance in 1923 as follows:

> At times one sees a strange behavior by bees that have returned home from a sugar water feeder or other goal. It is as if they had suddenly acquired the disease St. Vitus's dance [chorea]. While they run about the combs in an irregular manner and with slow tempo, their bodies, as a result of quivering movements of the legs, constantly make trembling movements forward and backward, and right and left. During this process they move about on four legs, with the forelegs, themselves trembling and shaking, held aloft approximately in the position in which a begging dog holds its forepaws. If they have brought in sugar water . . . often [they] will retain it until they have quieted down. The duration of this "tremble dance" is quite variable. I have seen instances where the phenomenon has died away after three to four minutes, then the bee appeared normal again and flew out of the hive. Usually, however, this dance lasts much longer and three times I have observed a bee tremble on the combs without interruption for three quarters of an hour.

The "strange behavior" described by Karl von Frisch is depicted in Figure 15.1. We see that a bee performing the tremble dance moves her body in three ways, and all at the same time: (1) *vibrational*—she steadily jiggles her body from side to side; (2) *rotational*—every second or so, she changes the direction that her head is pointing; and (3) *translational*—she walks slowly across the comb. The first time you see a bee walking around on a comb performing the tremble dance, you might suspect that there is something seriously wrong with her, for it does indeed look like her nervous system is malfunctioning, perhaps from pesticide poisoning.

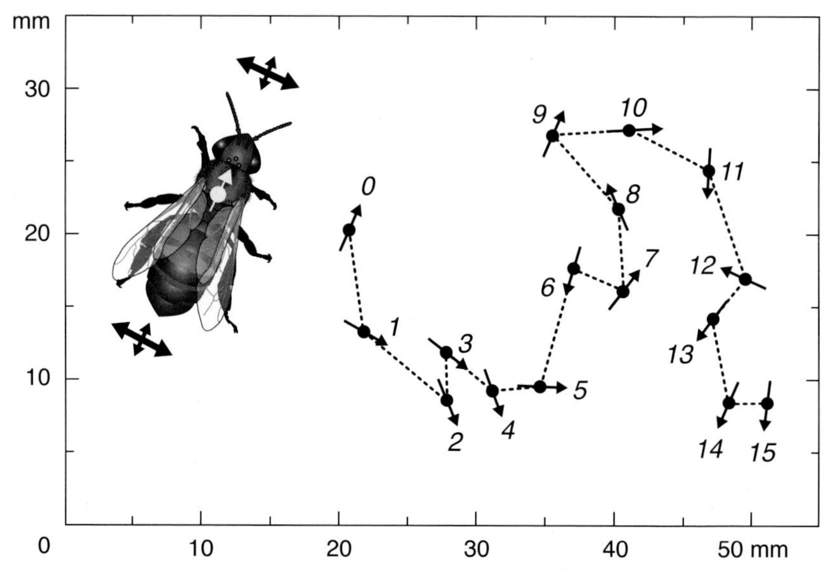

FIG. 15.1. A 15-second record of a worker bee's behavior while she performed the tremble dance on a comb inside an observation hive. The drawing on the left illustrates the eye-catching side-to-side movements (trembles) of a tremble dancer's body, while the track on the right shows how this bee walked slowly across the vertical comb and rotated her body-axis by 30°–90° every second or so. Note: 50 millimeters is approximately 2 inches.

It turns out, however, that actually she is producing an important signal that helps her colony take full advantage of a strong nectar flow.

The message of the tremble dance always puzzled Karl von Frisch, for although it—like the waggle dance—seemed to be a communication signal, he never discovered what elicits it, what its purpose is, or how it affects other bees in the hive. So, in 1923, he suggested that the tremble dance gives the other bees no information. Furthermore, in the 1960s, when he wrote his masterwork, *The Dance Language and Orientation of Bees*, he commented on the tremble dance as follows: "I think it tells the other bees nothing . . . and perhaps it is comparable to the condition that Florey has described as a neurosis." I remember first reading about this puzzling behavior in October 1974. This is when I was starting my Ph.D. studies and was reading carefully, from cover to cover, Karl von Frisch's 566-page magnum opus. Back then, I accepted his interpretation of this odd behavior, for he was the expert on these matters.

Thirteen years later, I discovered that the strange-looking tremble dance is not a symptom of a worker bee suffering a neurosis. Instead, it is a signal that plays an important role in the nectar collection process of a honey bee colony. As I will explain shortly, this discovery came as a surprise on 17 July 1987, at the Cranberry Lake Biological Station (CLBS), while I was conducting an experiment to analyze the phenomenon discussed in the previous chapter: a colony's nectar foragers are less choosy about the flower patches they advertise with waggle dances when their colony's rate of nectar intake is low than when it is high. In other words, a colony's nectar foragers behave as if they follow the human proverb "Beggars can't be choosers." In the summer of 1987, my aim was to figure out how nectar foragers know whether their colony's rate of nectar intake is low or is high.

I began by testing the hypothesis that a nectar forager can sense her colony's level of nectar intake by paying attention to how long she has to search to find a nectar receiver when she returns to the hive. Figure 15.2 shows the division of labor between workers who collect nectar and those who receive it and convert it to honey. As described in the previous chapter, I had learned in the summer of 1985 that there is a positive *correlation* between a colony's rate of nectar intake and how long its nectar foragers spend searching for nectar receivers when they get home. But, as we all know, a correlation does not imply causation. So, I needed to perform an experiment to test the hypothesis that a nectar forager takes her cue to adjust her waggle dance threshold from how long she has to search to find a nectar receiver. Performing this test required the help of the Cornell undergraduate students, Mary Eickwort and Oliver Habicht. Both wanted to see how one conducts rigorous experimental studies of animal behavior in the field. And both figured (correctly) that it would be fun to spend several weeks living at a remote biological station beside a beautiful lake in the Adirondack Mountains.

Our goal was to find out if it is a long in-hive search time to find a nectar receiver, or something else—such as a throng of nectar foragers dashing into the hive, or a strong aroma of fresh nectar inside the hive—that informs a colony's nectar foragers that their colony has a high rate

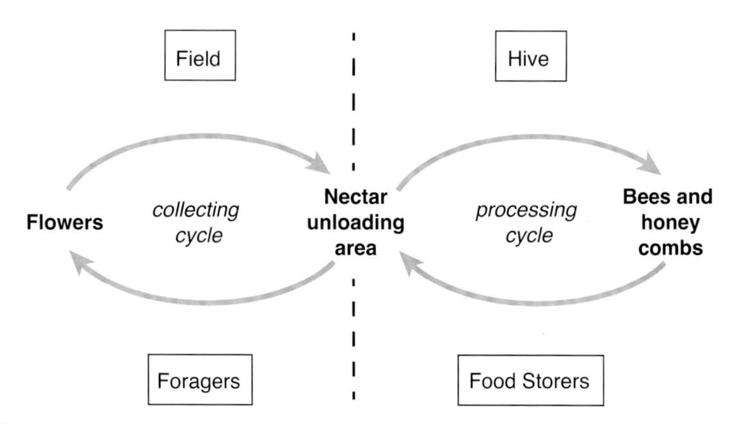

FIG. 15.2. The division of labor between nectar collectors and nectar receivers in honey bee colonies. The collecting cycle operates mainly in the field as foragers gather nectar from flowers, bring it home, unload it, and then dash back out to the flowers. The processing cycle takes place entirely in the hive. Nectar receivers unload nectar collectors, usually near the hive entrance. The nectar receivers share some of the fresh nectar with hungry nestmates; they store the rest in the colony's honey combs.

of nectar intake. To try to reach this goal, we performed an experiment that involved two colonies; one was a treatment colony and the other was a control colony. (The control colony provided a check for possible confounding effects of weather on the bees' behavior.) Each colony had a population of approximately 4,000 bees, and each was housed in a two-frame observation hive. Both colonies were positioned in the same place: outside one of the classroom buildings at the CLBS. But only one colony received the treatment that increased the search times experienced by its nectar foragers when they got home. This treatment was the removal of most of the colony's nectar receivers. In effect, we closed most of the "checkout lines" for the nectar foragers in the treatment colony. Then we watched to see if this manipulation discouraged the nectar foragers in this colony from performing waggle dances.

I must stress that our "trick" of removing most of the nectar receivers in the treatment colony increased the search times of the nectar collectors in this colony, but *it did not increase* the level of either the forager traffic into this colony's hive or the aroma of fresh nectar inside its hive. So neither a high forager traffic level nor a high nectar aroma level could

FIG. 15.3. Oliver Habicht tending the sugar-water feeder (sitting on the small table) that was visited by foragers from the treatment colony.

underlie whatever changes we might see in the behavior of the nectar foragers in the treatment colony.

We began our experiment by training a small number of foragers from each colony to visit one of two feeding stations (Fig. 15.3). Each station sat at the same distance (1,400 feet/420 meters) from the hives, but the stations were positioned in different directions from the hives (approximately southwest and south), so the two feeding stations sat 980 feet (300 meters) apart. Each colony's feeding station was the only strong source of "nectar" for this colony. Both feeders provided a rich (65% by weight) sucrose solution *ad libitum*.

The experiment unfolded over a four-day stretch of good weather: 14, 15, 16, and 17 July 1987. On each day, we allowed 15 bees (each one labeled for individual ID) from each colony to forage at their colony's sugar-water feeder. These 30 bees were eager to gather the sugar syrup we provided. I sat beside the observation hive that housed the experimental colony. When one of this colony's labeled foragers arrived home and scrambled into the observation hive, I tracked her and recorded how long she searched to find a nectar receiver and whether or not she performed a waggle dance. On 14 July, between 8:16 in the morning and 1:02 in the afternoon, I watched 68 return visits to the hive by the 15 labeled forager bees in *the treatment colony*. (These 15 bees made more than 68 return visits total, but I could follow only one bee at a time, so I watched only a fraction of their return visits.) I saw that their search times were short: on average, just 11 seconds. I also saw that these bees performed waggle dances during 73 percent of their returns to their hive. These dances recruited some of their hive mates to visit their feeder, but the recruited bees (recognized as such because they were not labeled) were captured and put in a cage, so the sugar-water feeder never got crowded. We released these bees after we shut down the feeding station for the day. The 15 bees in the control colony also recruited bees to their feeder, and these recruits, too, were caged temporarily to prevent the control colony's feeding station from becoming overcrowded. We needed the conditions at our two feeders to be as stable as possible—visited by just 15 bees each—throughout each day of the experiment.

Over the next two days, 15 and 16 July, we continued to allow 15 labeled foragers in each colony to make collecting trips to their colony's sugar-water feeder. Also, Oliver and Mary continued to temporarily "jail" the recruited bees that arrived at their feeders. But while they were doing this, I was no longer watching the behavior of the labeled bees in the treatment colony when they came into their observation hive. Instead, I was busy working on removing as many of the nectar-receiver bees as possible from the treatment colony. To do this, I spent each day daubing a dot of lavender-color paint on the thorax of every bee that I saw unloading any of the 15 bees bringing home sugar syrup from the treatment colony's feeder. (I had removed the glass from one side of the observation hive and replaced it with a screen of tulle through which I could insert a tiny paintbrush.) Then, at the end of both 15 and 16 July, I opened the treatment colony's hive and I plucked from its combs all the lavender-dotted bees (i.e., many of the nectar receivers in this colony) and put them in a cage with a sugar-water feeder. I needed these bees to stay alive and healthy. Over these two days (15 and 16 July), I labeled and removed 482 (about 12 percent) of the bees in the treatment colony.

On the morning of the fourth day (17 July), the weather was sunny and warm—hurray!—and I sensed that that this tricky experiment might bear fruit. At 8:05, Mary and Oliver reloaded their feeding stations and I began watching the bees in the observation hive, to see how the 15 nectar collectors that were visiting Oliver's feeder would behave when they returned home. Relative to the first morning of this experiment (14 July), would they find it harder to find nectar receivers and would they be less motivated to perform waggle dances, even though natural forage remained sparse and their feeder still provided rich sugar syrup? It was hard to be patient. The day started sunny but cool (61°F/16°C), so it was not until 8:55 that the labeled nectar collectors in the treatment and control colonies resumed their work of bringing home "nectar" from their separate feeders.

As soon as these bees resumed their work, I began mine. I repeated what I had done on 14 July. Specifically, I recorded two crucial things for each labeled bee (nectar collector) in the treatment colony when she entered the observation hive: (1) how long it took her to find a nectar receiver, and (2) whether or not she performed a waggle dance. Meanwhile, Oliver and Mary recorded how many *recruits* had arrived (and been captured) at their feeding stations in 15-minute intervals. (Oliver and Mary also reported these counts to me by walkie-talkie.) Soon, it became clear that my removal of nectar receivers from the treatment colony was having strong effects on this colony's nectar collectors. Now, when these bees returned home, I saw that they searched twice as long as before to find a nectar receiver (average, 21 seconds instead of 11 seconds), and that they performed almost no waggle dances! Three days before, on 14 July, I had seen 11 of these 15 bees perform waggle dances with great excitement, and Oliver had captured 23 bees per hour recruited to this colony's feeder. But now, on 17 July, I saw that only 1 of the 15 bees visiting the treatment colony's feeder performed waggle dances, and she did so sluggishly. Likewise, Oliver recorded only 3 bees per hour recruited to this colony's feeder. At the same time, it was clear that the 15 bees visiting the control colony's feeder must be dancing wildly inside their hive. Mary reported that lots of new bees (more than 40) had appeared at her feeder in the first hour, just as they had back on 14 July. This observation was important, for it showed that there was nothing about the weather that was causing the weak dancing by the bees visiting the treatment colony's feeder.

Watching at the observation hive that housed the treatment colony, I saw why its 15 labeled nectar collectors were no longer dancing excitedly: they were no longer getting "mobbed" by nectar receivers when they got home. Instead, they were experiencing difficulty finding bees willing to receive their loads of sugar syrup. Four times, I saw a labeled nectar collector give up looking for a nectar receiver and regurgitate her load of sugar syrup straight into a cell. I was astonished. Never before had I seen a nectar collector do this. *But what really gave me a mental jolt was seeing that 3 of my 15 labeled nectar collectors (20 percent) started to perform tremble dances shortly after they entered the observation hive.* Cool!!! I wondered, were these three bees issuing calls for more bees in their colony to function as nectar receivers? It sure looked like it.

That evening, when I entered into my notebook my "Summary and Conclusions" for the day, I wrote the following: "Noticed trembling dances by returning foragers. These bees did this after returning w/ nectar but had a hard time getting unloaded. . . . Is trembling a signal to increase the number of [nectar] receivers?" I knew we had stumbled on something important. In 1989, I described our findings in detail in a 19-page paper titled "Social foraging in honey bees: how nectar foragers assess their colony's nutritional status." Despite its dry title, this paper is one of my favorite scientific reports, because our two-colony experiment showed us how a nectar forager tunes her motivation to produce waggle dances (or tremble dances) by noting whether it is easy (or hard) to find a bee keen to receive her nectar.

In the summer of 1991, I returned to the Cranberry Lake Biological Station to perform an experiment designed to be a test of my hypothesis of what stimulates a nectar forager to perform a tremble dance: she returns home with a fine load of nectar, but then has difficulty finding a hive mate willing to receive her load. My test again involved increasing the unloading difficulty that a nectar forager experiences, and then seeing how she behaves. But unlike the experiment in 1989, it was based on mimicking a natural cause of increased difficulty in unloading: a surge in a colony's rate of nectar collection at the start of a nectar flow. When studying animal behavior, it is generally best to design your experiments so that your test subjects are presented with natural situations, for this helps ensure that your findings are relevant to the lives of the animals you are studying.

I was accompanied in 1991 by two Cornell undergraduate students, Erica van Etten and Timothy Judd, and by one colony of some 4,000 honey bees living in an observation hive. We trained foragers from this colony to visit two sugar-water feeders that were positioned 1,150 feet (350 meters) north and south of the hive. Each feeder was filled with a rich (65%) sucrose solution. Erica and Tim labeled the bees visiting their feeders with tiny paint marks that made these bees individually identifiable and that showed me at a glance (as I made observations at the hive) which feeder each bee was working. It was essential that I could distinguish the two groups of bees working the two feeders, because I wanted to gather data only from the bees visiting the north feeder.

Here is how the experiment worked. As in the 1987 experiment, we began by allowing only a modest number of bees (15) to make trips to each feeder; excess bees were caught and temporarily caged. But unlike in the 1987 experiment, we did not proceed to make it harder for the focal bees to get unloaded by removing many of their colony's nectar receivers, something that does not happen in nature. Instead, we increased the unloading difficulties of the focal bees (those visiting the north feeder) by increasing the number of bees collecting sugar water from the south feeder. This mimicked a natural rise in a colony's nectar intake, as happens when flowers offering nectar become more abundant. Figure 15.4

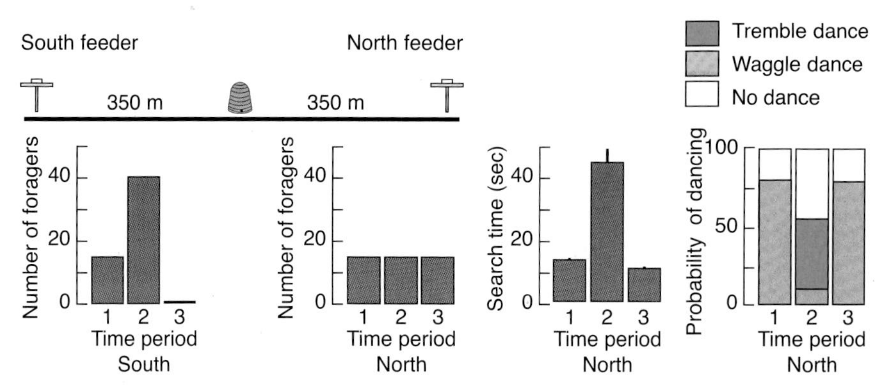

FIG. 15.4. Experimental setup and results of a test of the hypothesis that a lengthy search in the hive to find a nectar-receiver bee stimulates a nectar-collector bee to perform a tremble dance. Time periods: 1 = 11:00–12:00 a.m., 2 = 1:30–2:15 p.m, 3 = 2:30–3:00 p.m. Both feeders provided the same rich (65%) sugar solution throughout the experiment. The traffic level of the nectar foragers in this colony was adjusted by changing the number of bees (0–40) that made trips to and from the south feeder. Data were taken on the in-hive search times and dance behaviors of the 15 foragers that made trips to the north feeder. This figure shows that when the north-feeder bees experienced long search times (during time period 2) to find nectar-receiver bees, they switched from producing waggle dances to producing tremble dances.

shows the layout, the logic, and the outcome of the first trial of this experiment.

This trial began at 11:00 a.m. on the morning of 10 July 1991. Erica and Tim, who were tending the two feeding stations, loaded them with a concentrated (64%, by weight) sucrose solution, and I began watching the behavior of the foragers from the north feeder when they returned to the observation hive. There were two things about the behavior of these bees that I watched most closely: (1) how long a bee had to search to find a nectar-receiver bee, and (2) whether or not she performed a dance (either waggle or tremble).

Figure 15.4 shows what I saw. *In time period 1*, when only 15 bees were allowed to forage from each feeder (all recruits to the feeders were captured, so the colony's "nectar" intake stayed low), the bees from the north feeder experienced short search times for receivers, and these north-feeder bees performed only waggle dances in the hive. Then, *in time period 2*, when the assistant at the south feeder (Tim) stopped capturing the bees recruited to this feeder, so the total number of bees bringing home "nectar" nearly doubled (55 bees total instead of 30), the bees from the north feeder behaved very differently in the hive *even though nothing had changed at their feeder!* Observations made from 1:30 p.m. to 2:15 p.m. revealed that the north-feeder bees had much longer search times than before (45 seconds, on average, instead of 15 seconds) and they performed mostly tremble dances rather than waggle dances. Wow! Finally, *in time period 3*, when the south feeder was shut off so the colony's "nectar" intake dropped quickly to a low level, the north-feeder bees again experienced short search times and they again performed only waggle dances.

We performed a second round of this experiment the following day, and we got results that were identical to those from the first trial. This experiment made it clear that if a forager returns home from a first-rate nectar source and finds that she must search at length to find a nectar receiver, then she will perform a tremble dance. So, at last, the decades-old mystery of what elicits the tremble dance was solved!

Figure 15.5 shows what further work revealed about the relationship between the search time to find a nectar receiver and the kind of dance that a nectar forager performs when she gets home from visiting a desirable source of nectar. If her search time is below 20 seconds, then she is likely to perform a waggle dance, but if her search time is above 50 seconds, then she is likely to perform a tremble dance. And if her search time is in between, then she is disinclined to perform either type of dance, for she knows that her colony needs neither more nectar foragers nor more nectar receivers. You might say that her colony, which is a kind of honey factory, is humming along just fine.

Postscript: Several years later, in 2000, when a colleague, Dr. Susanne Kühnholz, and I were interviewing Professor Martin Lindauer in preparation for writing his biography, he told us that Karl von Frisch always suspected that the tremble dance is an important signal—not a symptom

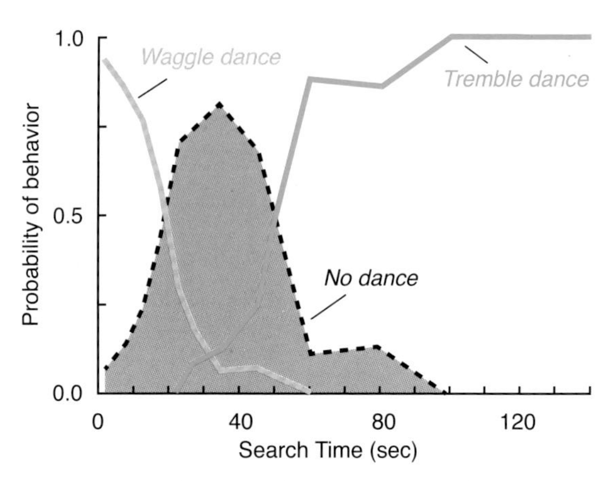

FIG. 15.5. Dance behavior as a function of the in-hive search time for foragers visiting a rich "nectar" source. Data were gathered from the 15 bees visiting the north feeder, as shown in Fig. 15.4. Their search times were varied by changing the number of bees bringing in "nectar" from a second feeder, south of the hive. The only changes underlying the switch from waggle dancing to tremble dancing were those experienced inside the hive: search time and perhaps other variables of the nectar-unloading experience.

of neurosis—and that he (KvF) had once told him (ML) that he would award a prize to whomever succeeded in deciphering the message of the tremble dance. Alas, Karl von Frisch died in the summer of 1982 . . . nine years before the summer of 1991, when my helpers and I performed the experiment that solved this long-standing mystery.

CHAPTER 16

Two Recruitment Dances, or Just One?

Most scientific discoveries have been made by researchers performing carefully planned tests of specific hypotheses about how nature works. However, some of the most beautiful discoveries have been made by chance—that is, by somebody noticing something unexpected in the course of his or her studies. A good example of a chance discovery is Sir Alexander Fleming's observation that when his petri dishes holding colonies of *Staphylococcus* bacteria became contaminated with a mold, the bacteria did not grow around the mold. It turned out that this mold was a rare strain of the fungus *Penicillium rubens*, and that it secreted a substance—eventually named penicillin—that inhibited bacterial growth. This was a momentous find, for it revealed the existence of antibiotics in nature, enabling their development into medicines for infections. Another shining example of a chance discovery is Karl von Frisch's observation that when a worker honey bee finds a rich food source, she can inform her nestmates about it by performing a dance. And this, too, was a momentous find. It started a line of investigations that revealed that honey bees possess a communication ability long regarded as unique to human beings: the ability to direct groupmates to an important location by providing information about its direction and distance, rather than by leaving a trail or leading others to the site.

Karl von Frisch began his studies of honey bees in the spring of 1911. He was then 24 years old and was working as an assistant professor in the Zoological Institute of the University of Munich. Back then, this institute occupied a building near the center of Munich that had been a Jesuit monastery, so it had a picturesque garden courtyard. This quiet spot provided von Frisch with a good space for conducting many of his early investigations on the visual and olfactory abilities of worker honey bees (Fig. 16.1). He started these studies by testing honey bees for color vision. To do so, he made arrangements with a beekeeper to have a hive holding a small colony of bees installed in the courtyard. Then von Frisch set out a shallow glass dish (a "watchglass") that held some honey. It sat on a table atop a 6 × 6-inch (15 × 15-centimeter) square of bright-blue paper. He observed that his dish of honey might sit unnoticed for several hours, but that once one worker bee had discovered it and had gone home, it was not long before dozens more visited this food bonanza.

The next day, von Frisch tested the bees for color vision by setting out an array of 16 *empty* dishes. One dish sat on a fresh square of the same blue paper that he had used the day before, and all the other dishes sat on squares of various shades of gray paper. He found that most of the bees examined just one dish . . . the one atop the blue square! In one test, for example, 406 bees went to the dish on the blue square, while only 13 bees in total visited the dishes on the gray squares. (*Technical note: each bee received a dot of yellow paint when she landed on a dish, so each bee was counted just once.*) This showed that the bees were capable of recognizing the blue square by its hue rather than by its brightness. Repeating this two-stage (training and testing) experiment with orange, yellow, green, and violet squares of paper gave results similar to the experiments with blue squares. If, however, the dish sat atop a red square of paper on the training day, then on the testing day the bees landed on the red square *and* on the squares of the darker shades of grey and black. This revealed that worker bees are blind to red.

Von Frisch continued his studies of bee color vision and odor discrimination for several more years. In the spring of 1919, however, he turned

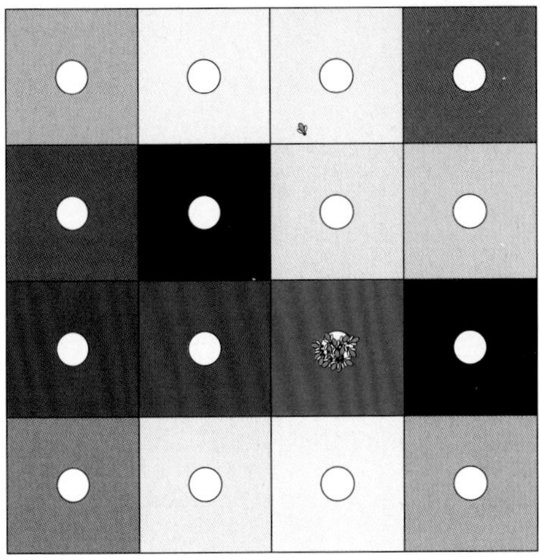

FIG. 16.1. *Top*: Karl von Frisch performing a study of *odor* discrimination by worker bees in the courtyard of the old Munich Zoological Institute. He has set up an array of eight cardboard boxes and is training bees to look for sugar water in only the box scented inside with jasmine. *Bottom*: Snapshot of the outcome of a test by von Frisch of *color* discrimination by honey bees. On day 1, von Frisch let bees collect honey from a glass dish that sat atop a square of blue paper on an otherwise empty table. On day 2, he set out (on the same table) a new square of blue paper amidst 15 squares of grey paper of various shades. Atop each square he placed a clean and empty glass dish. When bees arrived at the test array seeking more honey, they landed almost exclusively on the empty dish atop the square of blue paper. This showed that worker bees see blue as a color, not just as a shade of grey. Further tests showed that the blue square of paper and the grey squares of paper did not have distinct scents. Photograph courtesy of Bayerische Staatsbibliothek München/ Bildarchiv.

his attention to something mysterious that he had seen while conducting his studies of the bees' color vision. He had noticed that bees that had previously collected honey from a feeding dish would keep returning to it from time to time, even when the dish was empty. He had also noticed that when he reloaded the feeding dish and then saw one bee arrive there, load up, and fly home, it was only a few minutes before other bees appeared at the dish. Clearly, the bee that had found the dish refilled with honey had shared the news of her discovery. But how she had done this was not at all clear.

To investigate this mystery, von Frisch had his beekeeper friend bring to the Zoological Institute a small bee colony that was housed in a thin, glass-walled hive (a small version of what is shown in Figure 12.1). He positioned this hive on a table just inside an open window in a ground-floor room. Von Frisch then placed a small dish of sugar water outside the window, on a table in the courtyard. He waited for this food to be discovered, and then he labeled—with a dot of red paint on the thorax—each bee that visited his feeding station. After labeling about 30 bees, he shut down the sugar-water feeder for a few hours . . . long enough for the labeled bees to lose most of their interest in it.

When von Frisch saw the red-dotted bees standing quietly inside the observation hive, he again set out the dish of sugar water on the table in the courtyard. Then he waited there for one of his labeled bees to return and discover that again there was a bonanza food source sitting on the table. Eventually, a bee returned. And as soon as she landed at the feeding station and began to load up, von Frisch dashed indoors to his observation hive, for he wanted to watch this bee when she entered the hive. When she did, he saw something wonderful. In his words: "I could scarcely believe my eyes. She performed a round dance on the comb which greatly excited the marked foragers around her and caused them to fly back to the feeding table. This, I believe, was the most far-reaching observation of my life." It was clear to von Frisch that these round dances were calls to action.

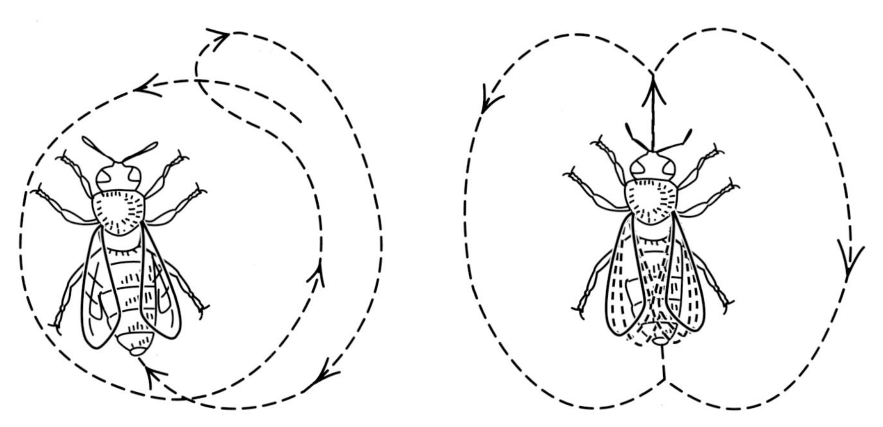

FIG. 16.2. Replicas of the drawings of the round dance (left) and the tail-wagging (waggle) dance (right) that are found in Karl von Frisch's first detailed report of his studies on the language of honey bees.

While doing this experiment, von Frisch noticed another form of dance behavior performed by worker bees in his observation hive. He called it the "tail-wagging dance" (in German, *der Schwänzeltanz*). Many of these "tail-wagging" bees bore loads of pollen on their hind legs. At this point, Karl von Frisch made a major error of thinking that honey bees have two distinct recruitment dances: round dances to announce rich sources of *nectar*, and tail-wagging dances (i.e., waggle dances) to announce good sources of *pollen* (Fig. 16.2). In 1920, he wrote, "It seems reasonable to see the two types of dance as different expressions of the bee language, of which the first signifies an ample nectar supply while the other means a good pollen source." [My translation from the German.]

With the benefit of hindsight, we can see that by setting up his sugar-water feeder in the courtyard of the Zoological Institute, thus only a few yards (meters) from his observation hive, von Frisch unintentionally created a strong correlation between food-source *distance* (near vs. far) and food-source *type* (sugar water vs. pollen). His bees had a nearby source of "nectar," but they had to fly off to flower gardens across Munich to get their pollen. In 1923, he described his findings on the dances of honey bees in his scientific monograph *Über die "Sprache" der Bienen* (*On the "Language" of Bees*).

Two decades passed before Karl von Frisch discovered that he had misinterpreted the two forms of the bees' dances. This discovery came in the summer of 1944. On July 12 that year, Munich suffered a heavy bombing raid that destroyed von Frisch's house. The next day, another big raid inflicted heavy damage on the Zoological Institute. Because von Frisch had no home left in Munich, and he could no longer work (or live) at the Zoological Institute, he and many members of his laboratory group left Munich and moved 130 miles (215 kilometers) east to his family's summer home in the hamlet of Brunnwinkl, which sits beside the Wolfgangsee, a small lake in the Austrian Alps. (He describes this site in detail in his 1980 book *Fünf Häuser am See* [*Five Houses by the Lake*]). Here he and his students carried on with their studies. They also devoted much time to gardening to produce their food.

Von Frisch's realization that he had misinterpreted the bees' dances emerged from a study of the odor discrimination abilities of worker bees that one of his colleagues, Professor Ruth Beutler, was conducting that summer in Brunnwinkl. She wanted bees from her study colony to forage at a sugar-water feeder that was scented with thyme and was located far from the bees' hive. Karl von Frisch advised her to train bees to collect a rich sugar syrup scented with thyme from a feeder near the hive (33 feet/10 meters away) and to also set up an identical feeder far from the hive (1,640 feet/500 meters away). He figured that the recruited bees would soon appear at both feeders. Ruth Beutler followed his advice, but had no success. All the recruited bees appeared at the nearby feeder. Now Karl von Frisch began to wonder: Did the distance of the feeder influence the dancing of the bees? Do the bees have a "word" for distance in their dances?

To answer this question, Beutler and von Frisch performed the experiment shown in Figure 16.3. They installed a small colony in an observation hive and then they trained two groups of bees from this colony to forage at two sugar-water feeders, one 33 feet (10 meters) and the other 820 feet (250 meters) away. Both feeders lay in the same direction from

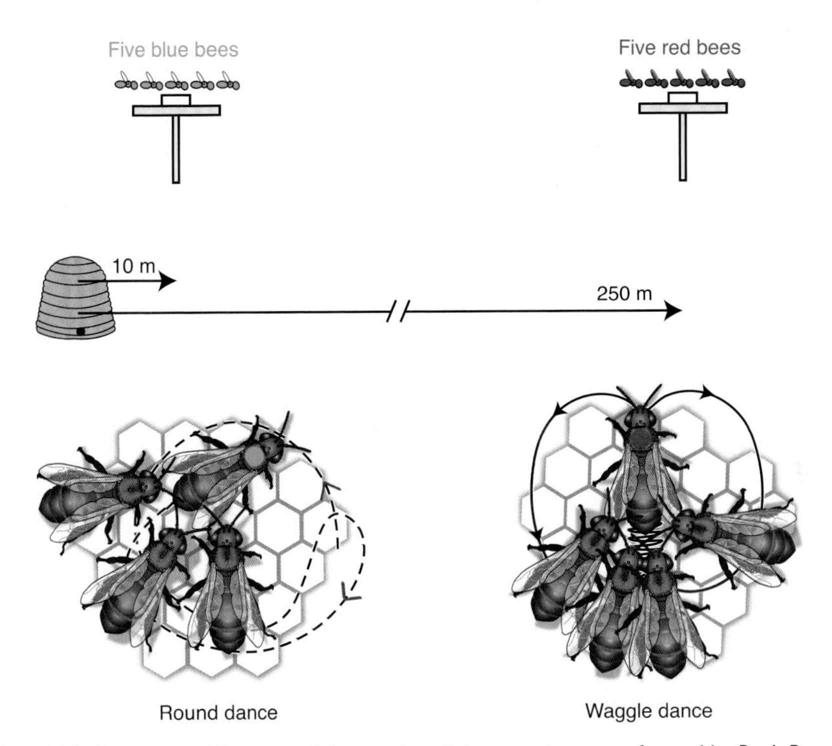

FIG. 16.3. Experimental layout, and the results, of the experiment performed by Ruth Beutler and Karl von Frisch to see if honey bees have a "word" for distance in their dances.

the hive. Once they had two sets of 5 bees—each one labeled with either a blue or a red paint mark for visiting one or the other feeder—they watched how these 10 bees behaved inside the observation hive. It was a breakthrough moment: the bees from the nearby feeder performed *round* dances, and the bees from the distant feeder performed *waggle* dances. It was now crystal clear that waggle dances were performed also by collectors of nectar, not just by collectors of pollen. Karl von Frisch realized that he had misinterpreted the correlation that he had seen between forage type (nectar vs. pollen) and dance form (round vs. waggle) when he began his studies of the bees' dances. He explains in Chapter 1 of his 1967 magnum opus, *The Dance Language and Orientation of Bees,* that, back in the 1910s and 1920s, he had seen in his observation hive waggle dances being performed by bees without pollen loads, but that he had interpreted the dances of these bees as those of pollen collectors that had already kicked off their loads.

Once Karl von Frisch realized he had made mistakes in interpreting the bees' dances, it did not take him long to decipher them correctly. Figure 16.4 shows what he figured out. A food source's *distance* is indicated by the duration of each waggle run in a dance performed to advertise the source. A food source's *direction* is indicated in a way that is not so obvious to us, but works well for the bees. They use the sun's azimuth (compass direction) as a reference direction, and they indicate the angle between the sun's azimuth and the food source's direction by pointing their waggle runs at this angle relative to straight up on the comb. So, for example, a bee advertises a food source that lies 40° to the right of the sun's direction by performing a dance in which each waggle run points 40° to the right of straight up, like what is shown in Figure 16.4. This was a beautiful, and truly amazing, discovery. It is no wonder that Karl von Frisch received the Nobel Prize in Physiology or Medicine for decoding the honey bee's dance language.

There is, however, one part of Karl von Frisch's work that recent studies have shown is not quite correct. It is his persistence in describing how honey bees recruit nest mates to rich food sources in terms of two distinct dances—round dances and waggle dances—even after he had learned that honey bees do *not* have distinct dances to announce nectar and pollen sources. For example, he still used his original, two-word framework to describe the honey bee's dance behavior when he wrote his 1967 masterwork. *The Dance Language and Orientation of Bees* has separate chapters devoted to round and waggle dances. Curiously, von Frisch does describe the smooth transition from round dances to waggle dances when a feeder's distance from the hive is increased from 33 feet (10 meters) to 330 feet (100 meters), but he does not stress the basic similarity in the forms of round dances and waggle dances.

In recent years, several teams of investigators have used tools—such as digital video recording and slow-motion playback—to look at round dances more closely than was possible when Karl von Frisch studied this

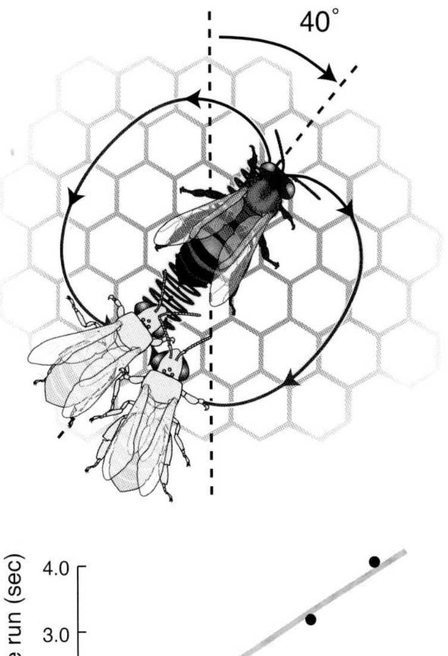

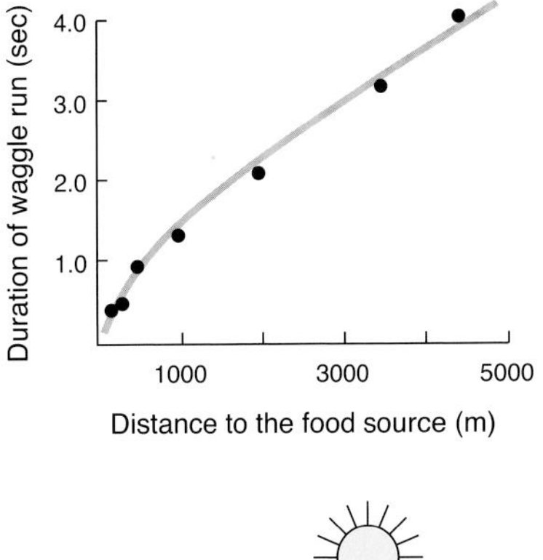

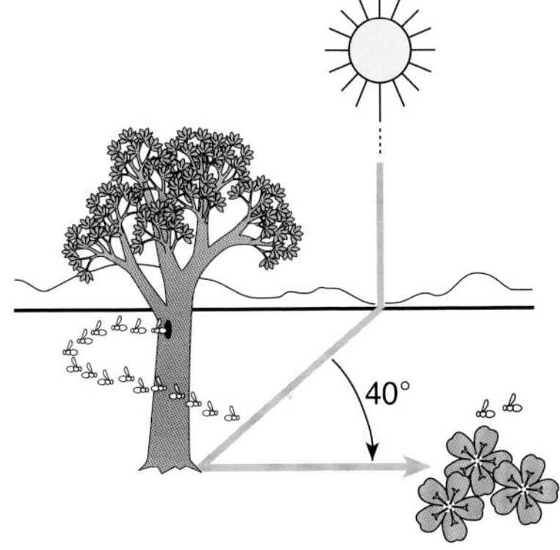

FIG. 16.4. How a worker honey bee encodes information about the distance and direction to a food source in her waggle dances. 1,000 meters = 0.62 miles.

behavior. These newer studies have reported the presence of brief waggle phases in the bees' recruitment dances even when the food source is only 33 feet (10 meters) from the hive. These studies did not, however, determine whether bees that have followed the dances for nearby food sources can *use* the location information that is encoded in these dances.

Karl von Frisch states in *The Dance Language and Orientation of Bees* that bees that have followed round dances "swarm out in all directions and examine the surroundings of the hive." But even in his most detailed description of round dances, published in 1923 (pp. 105–109), he provided only scanty evidence in support of this statement. So, in the summer of 2011, two of my students—Sean Griffin and Michael Smith—and I decided to take a careful look at this matter. Our goal was to perform an experiment that would answer the following question: Do bees that have followed round dances advertising a rich food source really "swarm out in all directions," or do they use the directional information that is encoded in these dances and search mainly in the direction of the rich food source?

We addressed this question by going to the Cranberry Lake Biological Station (CLBS), for I knew that the sparsity of natural forage here would make it easy to train foragers to sugar-water feeders. We worked with a colony living in a two-frame observation hive that was sheltered in a hut (see Fig. 16.5). We positioned this hut in the center of the large lawn at the CLBS. We then trained two groups of bees to forage simultaneously from two identical feeders (design shown in Figure 13.4) that sat in opposite directions (east and west) from the observation hive. One feeder provided a concentrated (65%) sucrose solution, so its bees *would* perform dances; the other provided a dilute (20%) sucrose solution, so its bees *would not* perform dances. We reasoned that if foragers use the directional information that is encoded in round dances, then more recruits should arrive at the richer feeder than at the poorer feeder. If, however, foragers do not use the directional information in round dances, then the recruits should arrive in similar numbers at the two feeders.

It was essential that our two feeders be identical in all ways except that one was, and one was not, advertised by dancing bees. So, both feeders were of the same design and were given the same scent, i.e., a drop of anise extract applied to a square of filter paper taped atop each feeder. Also, both feeders were visited by 10 bees, each of whom was labeled with paint marks for individual identification. Moreover, and critically,

FIG. 16.5. Layout at start of the test of the bees' ability to communicate direction information in the dances they perform to advertise nearby food sources. There were two feeding stations located 5 meters (16 feet) east and west of the orange hut, which housed the observation hive. A colony of some 3,500 bees occupied this hive. Author at left; Michael Smith at right.

every bee that visited our feeders did so with her Nasonov gland sealed shut. We had coated the rear abdominal terga (i.e., the dorsal surfaces of the two hindmost body segments) of each bee with several coats of shellac. This made it impossible for the bees working the richer feeder to mark it more strongly by releasing more assembly pheromone, which is produced in the Nasonov gland.

We started the experiment with our two feeders just 16 feet (5 meters) from the hive. One feeder was to the east and the other was to the west. One member of our team was stationed at each feeder, to capture the recruits (recognized as unlabeled bees) arriving at his feeder. Each recruit was captured in a plastic bag upon arrival. After capturing 54 recruits total at the feeders when they sat 16 feet (5 meters) from the hive, we moved the feeders simultaneously to locations 31 feet (10 meters) from the hive and in the same directions as before (east and west). After waiting 10 minutes, to give our labeled bees time to adjust to the feeders' new locations, we resumed counting the number of bees recruited to each

TABLE 16.1. Results of the test for directional recruitment by bees performing dances to advertise a feeder, when it was at various distances from their hive.

Distance in meters (feet)	*Recruits to feeder Rich (65% sucrose)*	*Poor (20% sucrose)*	*% recruits to the rich/poor feeder*
5 (16)	27	27	50/50
10 (33)	42	18	70/30
20 (66)	36	12	75/25
30 (99)	32	1	97/3
40 (132)	27	0	100/0
50 (165)	25	0	100/0

Note: The data from the two rounds of the test have been pooled. Both the rich feeder and the poor feeder were visited by 10 bees, but only the bees from the rich feeder performed dances. Each of the 20 bees had her Nasonov gland sealed shut. Also, every bee recruited to the feeders was captured, so the feeders never became crowded.

feeder. This procedure was repeated over the day, as the two feeders were moved farther and farther (to 20, 30, 40, and finally 50 meters) from the hive.

We conducted two trials of the experiment on two days. In the first trial, the advertised feeder was to the east and in the second trial the advertised feeder was to the west. The pooled results for both trials are shown in Table 16.1. You can see that when the two feeders (advertised and control) sat 16 feet (5 meters) from the hive, equal numbers of bees arrived at the two feeders. This shows that when we set up our feeders *very near the hive*, the words of Karl von Frisch about the bees' dances were correct: the bees that were excited by the dances did "swarm out in all directions and examine the surroundings of the hive." However, when we set up our two feeders *just slightly farther from the hive*—31 feet (10 meters)—we saw that many more recruits arrived at the advertised feeder than at the unadvertised feeder (42 vs. 18, thus 70 percent vs. 30 percent). Clearly, when the feeders were only 31 feet (10 meters) from the hive, the bees' dances provided the dance followers with some directional information. The evidence of the bees' dances providing directional information was even stronger when the two feeders were positioned 66 feet (20 meters) from the hive. Then 36 out of 48 (75 percent) of the recruits arrived at the advertised feeder. And when the feeders sat 99 feet (30 meters) or more from the hive, nearly all the recruits (97–100 percent) arrived at the feeder that was advertised by the bees' dances. Basically, every bee knew in what direction they should go to find the feeder. They did not swarm out in all directions.

I regret that it was not possible to share these results with Karl von Frisch. Alas, he died in 1982, thus 29 years before this experiment was performed. Even more, I regret that it was not possible to show Karl von Frisch slow-motion playbacks of the video recordings of worker bees performing dances to recruit nestmates to a feeding station just 66 feet (20 meters) from the hive. This would have enabled him to see, and hear, for himself the brief waggle runs that these bees produce during each circuit of their dances. If he had, then I believe that he would have said, "*Ja, es gibt nur eine Tanzform, der Schwänzeltanz.*" Translation: "Yes, there is only one dance form, the waggle dance."

CHAPTER 17

Movers and Shakers

Every September for more than thirty years, I moved two observation hives that were well stocked with honey bees from my off-campus laboratory to a room on the top floor of Stimson Hall, a handsome stone building in the center of the campus of Cornell University. I fastened each hive to a narrow table that supported both the hive and a long tunnel that led from the hive's entrance to the outdoors (Fig. 17.1). A week later, the 250 or so students enrolled in my department's popular course on animal behavior would come to Stimson Hall in small groups to do an "active learning" (i.e., laboratory) project with the colonies living in these glass-walled hives.

Each student observed the bees for as long as he or she wanted, chose a specific worker-bee behavior that piqued his or her curiosity, and then described this behavior in both words and a drawing. Each student also wrote four paragraphs to answer the following four questions about his/her chosen behavior: How does it work? How does it develop in the individual's lifetime? Why is it beneficial (i.e., adaptive)? What is its evolutionary history? These four questions—often referred to as "Tinbergen's four questions"—are the four complementary ways of explaining an animal's behavior. I had set up the observation hives to give the students the challenge of closely watching one honey bee behavior, then describing it in detail, and eventually thinking about it carefully by looking at it from all four perspectives. The only constraint on the students' choices of be-

FIG. 17.1. The two observation hives that for many years I moved onto the campus of Cornell University, for the "bee lab" exercise in my department's course on animal behavior. Hives like these were the most important tool in many of my studies of honey bee behavior.

havior was they could not use the famous waggle dance. This left, of course, countless other behaviors of worker bees, such as kicking loads of pollen off the hind legs while backed into a cell, fanning the wings while standing near the entrance, grooming another worker bee, licking the queen bee, receiving nectar from a forager that has just run into the hive, standing alert at the hive's entrance, and jamming a brown substance (propolis) into crevices.

A week later, the students turned in their bee-lab reports to be graded by me and another professor (and good friend), Dr. Paul W. Sherman. We both studied the behavior of animals—Paul was fascinated with ground squirrels, especially how they use their alarm calls—and we both wanted to give our students the experience of watching closely an animal's behavior. Although some students made minimalist drawings of the

FIG. 17.2. A worker bee producing the shaking signal. Having grasped a nestmate with her forelegs, she shakes her own body up and down (i.e., dorsoventrally) for 1 to 2 seconds, at a frequency of about 16 Hz. Then she breaks contact with the shaken bee, crawls across the comb, and usually shakes another bee. A bee can produce the shaking signal on 20 or more bees per minute, and she can continue doing so for several minutes. So, within one bout of signaling a bee can send this signal to dozens of her nestmates.

behaviors they had watched and wrote terse answers to the four questions about the behaviors they had described, others made precise drawings and wrote detailed answers to Tinbergen's four questions. One behavior that was chosen often, probably because it is eye-catching, is what I like to call the "shaking signal." It is depicted in Figure 17.2.

I was glad that many students chose this particular behavior. I felt this way because I knew that when a student saw a worker bee performing an eye-catching behavior that looks like a signal, he or she was confronted with a good mystery: What is the bee that is producing this signal telling the bees that are receiving this signal? In short, what message is the bee sending to her hive mates? We saw in the last two chapters how it took biologists, even gifted ones like Karl von Frisch, several decades to identify the messages that worker bees are sending when they perform waggle dances and tremble dances. We saw, too, that the messages of these dances were misunderstood until somebody figured out what the bees had experienced shortly before they produced their dances. The bees that performed waggle dances had gathered loads of rich food (nectar or pollen) at times when their colony *was* ready for a higher rate of nectar or pollen intake. The bees that performed tremble dances had gathered loads of rich nectar at times when their colony *was not* ready for a higher rate of nectar intake. The tremble dancing bees knew this was the state of their colony because they had to make lengthy searches to find bees willing to receive their loads of nectar.

The history of the study of the shaking signal shows again how difficult it can be to identify the message of a signal produced by a worker bee. This conspicuous behavior was described in the 1940s and early 1950s by at least six individuals in Germany, Russia, and the United States. All of these individuals suspected that the shaking behavior was a signal, but none of them identified its message or determined its function.

The behavior depicted in Figure 17.2 has been given various names, including dorso-ventral abdominal vibration ("D-VAV"), shaking signal, jerking dance, shaking dance, and vibratory (or vibration) dance. I think that "dorso-ventral abdominal vibration" is a misnomer because a bee producing this signal vibrates her entire body dorsoventrally, not just her abdomen. Also, I think that calling this behavior a "dance" is misleading because, unlike the waggle dance and the tremble dance, it is a very brief behavior (just 1–2 seconds long), so it does not look like a dance. I feel that the most appropriate of these various names is "shaking signal."

The first solid evidence about the message of a shaking signal came from two studies that focused on worker bees shaking their mother queen, not their fellow workers. One of these studies was conducted by M. Delia Allen, a young woman who worked in the Beekeeping Research Department of the North of Scotland College of Agriculture, near Aberdeen. Her studies of the occurrence and significance of the shaking signal earned her a Ph.D. from the University of London in 1958. She began her investigations in 1954, at the age of 25, with the aim of understanding how the workers in a colony prepare their mother queen for her departure in a swarm. Her first step was to establish a small colony headed by a year-old queen in a three-frame observation hive that was connected to the

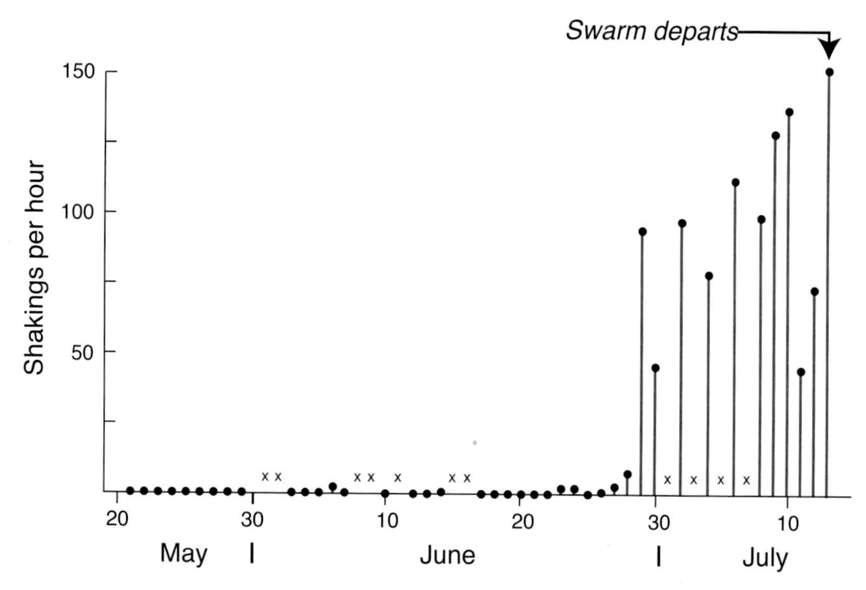

FIG. 17.3. Shaking of the queen by workers in a honey bee colony preparing to swarm. Queen cell building started on 23 June, and a few days later the workers began to shake the queen, evidently to prepare her for flight. The swarm departed on 13 July. The x's mark days when data were not collected.

outdoors by a tunnel, like what is shown in Figure 17.1 She then watched her colony, to see how the workers' behavior around their queen changed as the colony prepared to swarm. Her observations started on May 18 and continued through mid-July. She saw that her colony cast a swarm with its mother queen (a "prime swarm") on 7 July, and then cast multiple afterswarms (number not specified) between 8 and 18 July. She also saw that the shaking of the mother queen started on 8 June and became "almost continuous" by 16 June.

These pilot observations, together with further observations made in 1957, when Allen recorded in greater detail how a colony's queen was shaken as the colony prepared to swarm (Fig. 17.3), showed that the workers' shaking of their queen is a means of preparing her for flight. The numerous bouts of shaking caused the queen to walk around more than usual. This increased walking by the queen—together with decreased feeding of her and decreased egg laying by her—caused her to lose weight during this pre-swarm period. I estimate that a mated queen sheds about 25 percent of her body weight when a colony prepares to swarm, and so gets herself into trim for flying.

The conclusion of M. Delia Allen about the function of shaking signals directed at a queen matches that reached by another Ph.D. student, Eleonore Hammann, who studied at the Free University in Berlin in the 1950s. However, instead of looking at how worker bees shake their *mother* queen in preparation for swarming, Hammann investigated how worker bees behave toward a *young* queen as she prepares for her mating flights. Hammann observed that at first a colony's workers will gently feed, touch, and lick their new (unmated) queen, but that in time they will shake her—by grasping her with their legs and then shaking their bodies—and that this rougher treatment appears to cause her to leave the hive on her mating flights. Hammann also reported that these "attacks" cease when the queen begins her egg-laying. So, her findings also indicate that the message of the shaking signal is "Prepare for flight" or "Prepare for greater activity."

After M. Delia Allen and Eleonore Hammann did their studies in the 1950s, other biologists studied the shaking signal further, but with a focus on workers shaking workers (rather than queens), as depicted in Figure 17.2. Workers are the most common targets of the shaking signal, and here again the message of the shaking signal remained mysterious for many years. This situation changed in the 1980s when Stanley S. Schneider, now a professor at the University of North Carolina at Charlotte, looked closely at this behavior for his Ph.D. studies at the University of California, Davis. He found that worker bees functioning as foragers will produce this signal both before they have started their work, early in the morning, and after they have started their work. These findings suggested strongly that the shaking signal, when directed at workers, has the same message as when it is directed at a queen: "Prepare for greater activity." Schneider and colleagues also observed that forager-age workers (but not younger ones) that have received the shaking signal become more active in the hive and tend to move to the "dance floor." This is the region of the combs, near the hive entrance, where waggle dances are performed.

Work by James C. Nieh, a professor at the University of California, San Diego, has confirmed these findings. He, too, found that workers move faster after being shaken than before, and that after being shaken their walking is oriented toward the hive entrance (and thus the dance floor), whereas before being shaken their walking has a random orientation. All these studies made it clear that the shaking signal, when directed at workers, plays a role in priming these bees to work harder so that their colony makes good use of rich foraging opportunities. I sensed, though, that there was more to learn about this signal. In particular, I wondered: Are there particular experiences that will stimulate a forager bee to produce shaking signals?

The answer to this question began to emerge in the summer of 1994, when I worked at the Cranberry Lake Biological Station with two students from Germany, Susanne Kühnholz and Anja Weidenmüller. None of us was thinking about the shaking signal when we began our work. Our goal then was to get a better understanding of how honey bee colonies control their water collection. This was Susanne's research project for her *Diplomarbeit* (roughly the equivalent of a master's thesis) at the University of Würzburg, in Germany. One of the questions that we tackled in June that year was whether there are interactions between a colony's water collection and nectar collection operations. For example, if a colony's need for water shoots up, does this inhibit the colony's collection of nectar? Or do the processes of water collection and nectar collection function independently? To address these questions, we trained a group of 15 bees—each one labeled for individual identification with paint marks—to forage at a sugar-water feeder; then we stimulated the colony's water collection by heating its hive and saw whether this inhibited the colony's nectar collection. It did not. In Chapter 19, when we examine how a colony controls its water collection, we will see why we got this result.

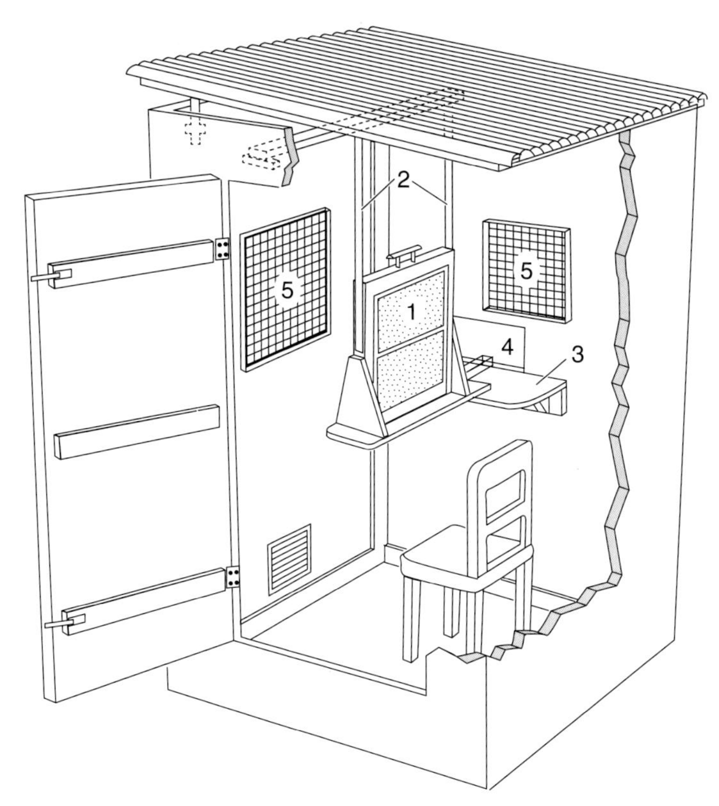

FIG. 17.4. The portable observation hive and hut: (1) Hive with transparent entrance tunnel, (2) threaded rods by which hive is suspended from a metal bar overhead, (3) shelf to which the hive is anchored, (4) window at the end of the entrance tunnel, (5) shuttered windows for light in addition to that provided by the translucent roof. Based on a drawing by Barrett Klein.

The sugar-water feeder that Susanne, Anja, and I set up for this work sat in a small clearing 1,150 feet (350 meters) south of the special hut (shown in Fig. 17.4) that sheltered the observation hive in which our study colony lived. Our feeder was the only significant source of "nectar" available to this colony (as discussed in Chapter 12). But because of vagaries in the weather, we did not fill our sugar-water feeder every day. Specifically, there were strings of days in June 1994 when the weather was so cool and rainy at the CLBS that the foragers of our study colony would have had difficulty getting out to our feeder, and at these times we left it empty. We did not want to tempt our labeled bees to make foraging trips that could be dangerous.

It was at the end of one of these spells of cool, wet weather that one of the bees visiting our sugar-water feeder provided us with a big clue about what exactly stimulates a nectar forager to produce shaking signals. The days of June 15 and 16 were chilly and rainy at the CLBS, so I did not set up our sugar-water feeder and our bees did not leave the observation hive. June 17, however, was warm and sunny, so I filled the feeder shortly after 8:00 a.m., settled into my lawn chair, and waited for the bees to resume their collecting trips to the feeder. It was my job to record when each forager (all were labeled for individual ID) arrived and when she departed. It was not long before one of these bees showed up. It was the bee bearing a dot of red paint on her thorax ("Red Thorax"). I said to myself, "Good morning, little friend." Then, I called (via walkie-talkie) to Susanne and Anja, who were sitting beside the observation hive, to alert them to keep a sharp lookout for Red Thorax and to tell me if she performed a waggle dance when she got back inside the hive. I certainly expected her to do so, for the feeder was loaded with a rich sugar syrup and the colony's rate of nectar intake was rock bottom. A few minutes later, Anja reported that when Red Thorax came into the observation hive, she had scrambled onto the comb just inside the hive's entrance as usual, but then had behaved strangely. Rather than perform a waggle dance, she had dashed around grabbing other bees (one at a time) and then shaking her body up and down (i.e., dorso-ventrally) for a second or so. Cool! I couldn't see Red Thorax, but I could tell that she was producing shaking signals. Susanne and Anja also reported that the behavior of Red Thorax was extremely conspicuous, because every other bee standing near the hive's entrance—the colony's other foragers—stood almost motionless, just as they had throughout the last two days of foul weather. This made me wonder, was this bee producing shaking signals to rouse her fellow foragers to return to work? Or was there another explanation for this bee's surprise outburst of shaking signals?

We talked things over that evening, and resolved to examine more closely the phenomenon that had unfolded that morning, to get a sharper picture of what stimulates a worker bee to produce shaking signals. We did so several weeks later, in July 1994, by performing another experiment at the CLBS with the colony living in our observation hive. We allowed a group of 15 bees—each one labeled for individual ID with a color & number bee tag, as shown in Figure 12.4—to forage at a sugar-water feeder of the design shown in Figure 13.4. This feeder was set up as before, 1,150 feet (350 meters) south of the hive, in the clearing by the bridge over Sucker Brook, and again it was the only good source of "nectar" available to our colony. But now we held back intentionally from filling this feeder every day. From time to time, we left it empty for 2–4 days in a row. This was our way of testing our hunch that a sudden resurgence in nectar availability following a lengthy dearth causes a forager that has found a bonanza nectar source to produce shaking signals to stimulate other nectar foragers to get back to work. We predicted that if this hunch was correct, then we should find a tight association between when we refilled our sugar-water feeder (after leaving it empty for a few days) and when we saw our labeled foragers produce shaking signals.

We performed the protocol four times of (1) turning our sugar-water feeder off for several days, then (2) turning it back on and seeing which bee discovered that it was again a first-rate food source, and finally (3) recording what this bee did when she got back inside our observation hive. Let's take a look at what we saw when we tracked the bee Red-30, who was the first bee to find the refilled feeder in one replicate of our experiment.

This particular bee's history was as follows. She emerged from her brood cell on 1 June 1994, and later that day she received her ID tag. Red-30 then worked inside and outside the hive without drawing our attention until she appeared at our sugar-water feeder late in the day on 16 July, hence when she was 45 days old. Despite her advanced age—in summer, 45 days is a respectable age for a worker bee—she foraged steadily at our feeder. Specifically, Red-30 visited it regularly whenever it was filled on 17, 18, 19, and 20 July. But when we left our feeder empty on 21 and 22 July (both were fair-weather days), she visited it only sporadically.

On the morning of Saturday, 23 July 1994, at 9:15 a.m., I refilled the feeder with sugar syrup. This ended a 64-hour period of "forage fasting" during which we had closed down the one good "nectar" source available to our study colony. The day started sunny and warm, but most of the bees in our study colony remained quiet, essentially motionless, inside the observation hive, probably because there had been little or no food for them to collect during the previous two days. But at 9:19 a.m., Red-30 roused herself and exited the hive. She arrived at the feeder at 9:20 a.m., and because she was the first bee to do so, she became our focal bee. She quickly loaded up on our sugar syrup, and at 9:22 a.m. she flew off. At 9:24 a.m., Susanne and Anja watched her scurry through the entrance tunnel and into the observation hive. Immediately upon climbing onto the hive's lower comb, Red-30 began to produce a string of shaking signals; she shook 33 bees during her first 92 seconds inside the hive. Next, she regurgitated, to two nectar receivers simultaneously, the load of sugar syrup she had collected. Then she quickly shook 16 more bees and dashed out of the hive. She was excited! For the next three hours, I recorded her comings and goings at the feeder, and Susanne and Anja recorded her doings in the hive, especially the signals (shaking signals and waggle runs) that she produced.

Figure 17.5 shows what Susanne and Anja recorded. Red-30 produced only shaking signals after each of her first 8 trips to the feeder; then she usually produced a mixture of shaking signals and waggle runs after her next 18 trips to the feeder; and finally she produced only waggle runs after each of her last 7 trips to the feeder, which I shut off at 12:30 p.m.. I did so because by then the sky had filled with dark clouds, the temperature had dropped, and we did not want to endanger Red-30 by tempting her to continue foraging during a thunderstorm. Thank goodness, when rain started coming down in sheets at 12:31 p.m., our star bee, Red-30, was safe inside the observation hive.

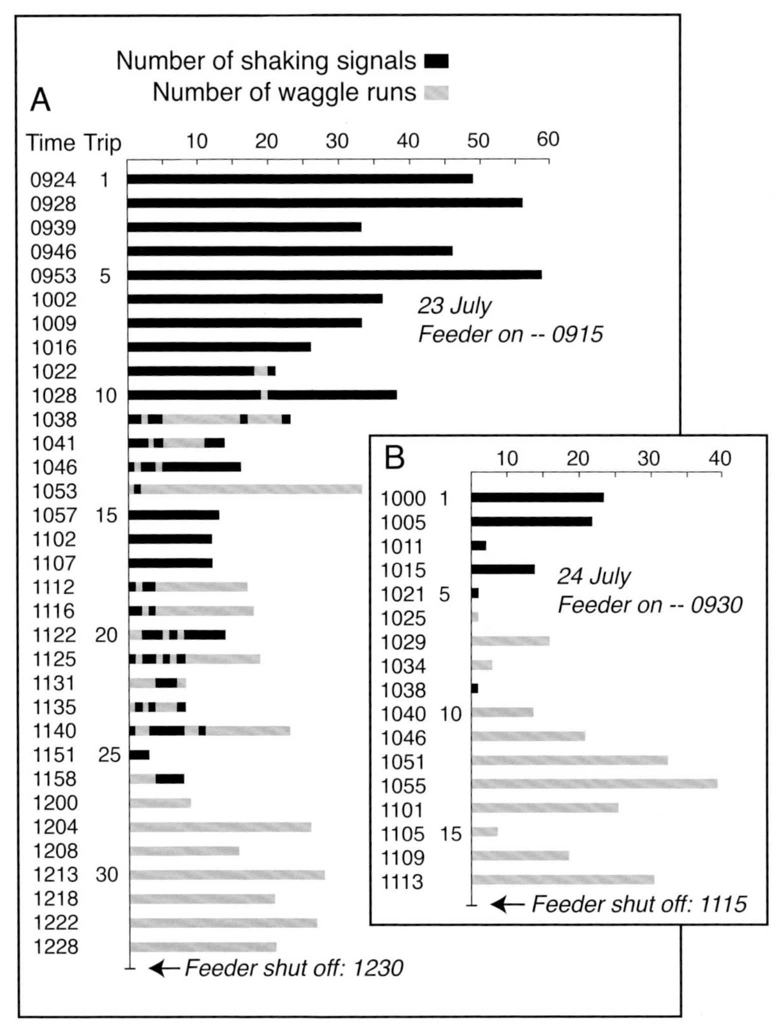

FIG. 17.5. Production of shaking signals and waggle runs by one forager bee, Red-30, on the mornings of 23 and 24 July 1994. On the two days preceding these observations (21 and 22 July), she had no foraging success because we had intentionally left the feeder empty. Thus, her discovery of sugar syrup at the feeder on the morning of 23 July followed a 2-day period of poor foraging, both by herself and by her entire colony.

The next day, 24 July, began cool but sunny. Red-30 resumed her work shortly after 10:00 a.m. and again she began by producing only shaking signals when she returned home from our feeder. But she did so with less resolution than on the previous day. On that day, she had shaken bees after her first 26 trips to our feeder, but on this (second) day, she did so after only her first 5 trips plus one more trip later on. Also, she shook fewer bees per return to the hive on the second day than on the first day. Her determination to produce waggle dances, however, had the same strength on both days: on average, she produced 21 waggle runs and 19 waggle runs per return to the hive on 23 and 24 July, respectively.

Thus, we saw on both days that our go-getter bee, Red-30, performed exclusively shaking signals when she made her first returns to the hive, evidently to rouse her sluggish hive mates. Then she shifted to mixing shaking signals and waggle runs. Eventually, she produced only waggle runs. Susanne and Anja also noticed on both days that by the time Red-30 had finished producing her shaking signals, the other forager bees in the colony had become active. Specifically, they were no longer standing motionless on the combs, and instead were crawling about, following waggle dances, and flying to and from the hive.

At this point, you may be wondering whether the immobile foragers that the bee Red-30 had shaken so excitedly were sound asleep and were getting wake-up calls. Back in 1994, we did not know for certain whether worker honey bees sleep, but now we know that they do, so I am sure that these immobile foragers were in a deep sleep (Fig. 17.6). Most of what we know about sleep in honey bees comes from the work of Barrett A. Klein, a professor at the University of Wisconsin-La Crosse. Barrett is both a gifted artist and a talented experimentalist. He started working with honey bees when he was an undergraduate student at Cornell University, where he majored in entomology, but also took courses of various sorts, including my department's course in animal behavior. He and I met during the honey bee laboratory in this course, and when I saw the precise drawings that he submitted for his lab report, I was dazzled! I knew I wanted to work with this talented and (I soon learned) exceptionally good-natured young man, so I was delighted when he signed up to help me perform experiments at the CLBS in 1992.

Barrett quickly learned how to do behavioral experiments with honey bees living in observation hives. And he began thinking about how he might work with honey bees to answer a question that had intrigued him

FIG. 17.6. Worker bees in sleep posture, hanging limply from the wooden frame of a comb in an observation hive. Sleeping bees are usually motionless, but can at times twitch their antennae or tarsi (feet). To distinguish a sleeping bee from a resting bee, one looks at her abdomen; the abdomen of a bee asleep pumps back and forth only occasionally, and in short bouts.

for years: Do insects sleep? He began to tackle this question with paper wasps, for his master's work at the University of Arizona, and then probed it further with honey bees for his Ph.D. work at the University of Texas at Austin. Laboratory-based studies that suggested that insects sleep had been performed already in Germany by Professor Walter Kaiser (studying honey bees) and in Switzerland by Professor Irene Tobler (studying cockroaches). Their studies set the stage for studying insect sleep under more natural conditions. Barrett first confirmed what Walter Kaiser had described in his studies of individual forager bees in the 1980s: the elderly (more than 16-day-old), forager bees in a colony have (like us) a 24-hour sleep-wake cycle. Barrett also found that the *elderly, forager bees* spend their nights largely asleep *outside of* cells that are away from the bustling brood-nest region of their colony's nest (Fig. 17.6). Meanwhile, the *young, nurse bees* sleep at various times of the day and night by taking naps *inside* empty cells that are within the central, brood-nest region of their home. Barrett then performed an ingenious experiment at the CLBS in 1994 that shows that getting good sleep at night helps the foragers in a honey bee colony function well the next day.

This experiment involved setting up two groups of forager bees—a *treatment group* of bees that had their sleep disrupted for a night, and a *control group* of bees that were allowed to snooze normally—and then comparing the precision of the waggle dances performed the next day by the bees in the two groups. To perform this experiment, Barrett glued a small disk, made either of steel or copper, atop the thorax of each of 50 foragers—25 for each type of disk. He had already trained all 50 bees to forage from a sugar-water feeder 1,090 yards (one kilometer) from their home, which was an observation hive. Then, all night long, and with a frequency of three times per minute, he slid past each window of the hive a tall strip of Plexiglas (mounted on rollers) that had powerful magnets screwed to it. This jostled the steel-tagged bees three times per minute, so they did not get good sleep, but it did not disturb the copper-tagged bees. The next morning, Barrett video recorded the waggle dances that the treatment (sleep-deprived) and the control (not sleep-deprived) foragers produced to advertise the sugar-water feeder that they were all visiting. Then a team of helpers—all of whom had no knowledge of the experiment's design—painstakingly analyzed Barrett's videos of the dancing bees. This work revealed that the variation in the angles of the waggle runs in a bee's dance (the indication of direction to the feeder, see Fig. 16.4) was greater for the sleep-deprived bees than for the well-rested bees. This result shows that forager bees, like you and me, function best after getting a good night's sleep.

In a further analysis of the video recordings made in that experiment, Barrett and his students examined 615 instances in which a bee followed a dance of either a steel-tagged (sleep-deprived) or a copper-tagged (control) bee. They found that bees that followed the (less precise) dances of the steel-tagged bees were likely to switch to following the dance of a different bee. This is an abnormal response. In contrast, the bees that followed the (more precise) dances of the copper-tagged bees were likely to dash out of the hive. This is a normal response. I like these results very much, for they show us that the bees themselves, not just the biologists equipped with video cameras, had sensed the dance imprecision that was caused by the sleep deprivation of some of the foragers.

There remains much to be learned about sleep in honey bees. Do a colony's queen and its drones need sleep, and if they do (which seems likely), then when and how do they sleep? Does the queen take "catnaps" around the clock, like the young workers? Do drones enjoy deep and restful slumbers at night, like the foragers? What is clear at this point is that the foragers in honey bee colonies, like ourselves, do need a good night's sleep to function properly. It is also clear that some of a colony's foragers, like some of us, are inclined to sleep in come morning. I hope that someday you will get a chance to peek inside a bee hive early on a fair-weather day, and see worker bees producing shaking signals. You can be sure that these shaker bees are delivering wake-up calls to their sleepy hive mates.

CHAPTER 18

Groom Me, Please

Good grooming is important for human beings, but it is indispensable for honey bees. These insects live crowded inside tree hollows, bee hives, and other tight spaces, so they must cope with a never-ending rain of wax bits, pollen grains, and other debris falling onto their bodies. If you install a sheet of sticky cardboard on the floor of a hive (beneath a screen), perhaps to measure a colony's level of infestation with mites, then you will see how much stuff falls down inside a colony's home in just a day or two . . . LOTS! The workers deal with this rain of "trash" by grooming both their own bodies ("autogrooming") and those of their nestmates ("allogrooming"). The queen and drones also spend much time grooming, but each of these individuals grooms just herself or himself. The queen grooms herself to spread the 9-ODA from her mandibular glands back over her body. A drone grooms himself to stay in top trim for flying. If you watch closely at a hive's entrance early in the afternoon on a sunny summer day, you will see drones carefully cleaning their antennae before they zoom off on their mating flights. Indeed, every bee in a colony gives her or his antennae special care, for these appendages are the seats of a bee's senses of olfaction and touch.

One of the fundamental truths in biology is that "form follows function." A beautiful example of this is the antenna cleaner structure that all honey bees—workers, queens, and drones—have on their forelegs. (Incidentally, all bees, not just honey bees, possess antenna cleaners on their

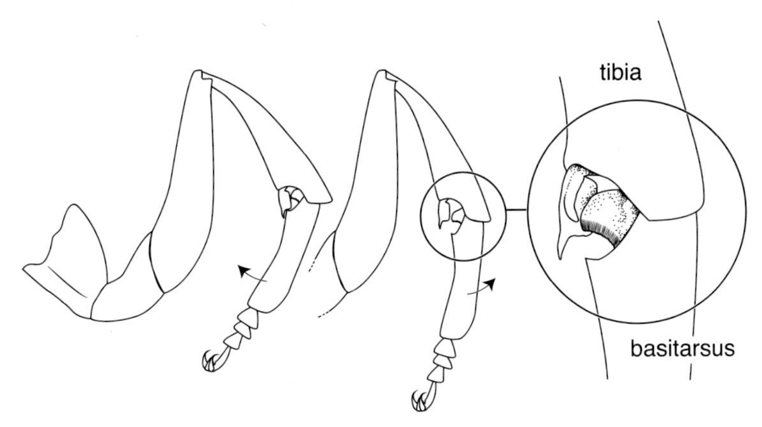

FIG. 18.1. *Left and Middle:* Rear view of the structure of the right foreleg of a worker honey bee, showing the antenna cleaner at the top (proximal end) of the basitarsus, in its closed and open positions. *Right:* Close-up view of the antenna cleaner notch on the basitarsus and of the scraper spur at the bottom (distal end) of the tibia.

forelegs.) A bee's antenna cleaner has two parts. The first is a semicircular notch located on the inner surface of the upper (proximal) end of the foreleg segment called the basitarsus (Fig. 18.1). This notch is lined with a comb of short, stiff hairs. The second structure is a large flattened spur which extends downward from the next higher foreleg segment, the tibia. This tibial spur has no muscles, but when the basitarsus is pulled toward the tibia, the spur closes the semicircular notch and functions like a rubber scraper. So, how does a worker bee use her pair of antenna cleaners? To clean her right antenna, for example, she raises her right foreleg and passes it over the right antenna so that the antennal base slips into the notch. She then flexes the tarsus so that the antenna is brought up against the spur. The antenna is now held securely in the cleaner. Finally, in the blink of an eye, she draws her antenna upward through the cleaner. The comb of fine hairs lining the notch cleans the antenna's outer surface, and the spur scrapes clean its inner surface.

I recommend watching this cleaning behavior the next time you have an opportunity to observe bees living in an observation hive, or to watch bees leaving a beekeeper's hive. In the latter setting, you will see that many bees pause on the landing board and clean their antennae just before they launch into flight. Devoting a few minutes to watching a bee skillfully swipe clean the outer segments of her (or his) antennae will give you a lasting memory.

There is also a second form of autogrooming by worker bees. It is the self-grooming behavior that every pollen forager performs to fill the pollen baskets on her hind legs. One especially nice feature of this grooming behavior is that it is easy to watch. It lasts for several seconds at a time and it takes place outside the hive, wherever honey bees are gathering pollen. Furthermore, we can watch it first thing in spring, when we are eager to see our insect friends again following months of separation forced on us by winter's cold. Dandelion flowers are excellent for this form of bee watching because they are common, and because when bees scrabble among the anthers of dandelion flowers they get dusted all over with pollen (Fig. 18.2). The bees then take pauses to brush the pollen off their bodies and tuck it into their pollen baskets. It is also easy to observe the pollen-packing process if you can find bees working on poppies (e.g., Flanders poppies, *Papaver rhoeas*). Their flowers have no nectar, so every bee working them is collecting pollen. Also, poppy flowers sit atop tall stems, so when bees are packing pollen in a patch of poppies, they are hovering a few feet (a meter or so) above the ground. It is easy to imagine that these bees are showing off their skills.

To watch the pollen-packing behavior, first find a worker bee that is hovering over pollen-bearing flowers and is well dusted with pollen. Then try to get close to her, say only a yard (meter) or so away. If you can do so, you should be able to see the sophisticated leg and tongue movements that she executes to load her pollen baskets. These movements are fast, so here is a guide to what you will see. *Step One*: She regurgitates a bit of nectar from her honey stomach onto her tongue, and then she strokes her tongue with her forelegs. This makes the brushes on her forelegs moist and sticky. *Step Two*: She sweeps the pollen off her antennae using the antennal cleaners on her forelegs, and she sweeps the pollen off her head and the anterior regions of her thorax using the sticky brushes on her forelegs. *Step Three*: She sweeps the pollen from her forelegs and the

Fig. 18.2. Pollen grains on the antennae, head, and legs of a worker bee that is foraging on dandelion (*Taraxacum officinale*) flowers. The hairs on her head and thorax (not shown) are finely branched and plumose (feathery), which makes them excellent for picking up and retaining pollen. The hairs on the abdomen of a worker bee are short and are not finely branched, so they pick up little pollen. Shown, too, is the structure of a bee's antennae: the rigid basal section (the scape) and the flexible distal section (the pedicel and the 10 segments of the flagellum).

rear part of her thorax using the brushes on her middle legs. She then moistens the loose pollen on these brushes with regurgitated nectar to make this pollen sticky. *Step Four*: She grasps each middle leg (one at a time) between the combs on the inner surfaces of both her hind legs and then she draws the middle leg forward. This leaves a mass of sticky pollen on the rows of stiff combs on the inside surface of each hind leg. *Step Five*: Once the combs on her hind legs are loaded with pollen, she shifts it to her "pollen baskets" with some more fancy "legwork."

The process of shifting the sticky pollen to the pollen baskets begins when the bee positions a rake structure—located at the bottom of the tibia of one hind leg (see Fig. 18.3)—against the top of the combs on her

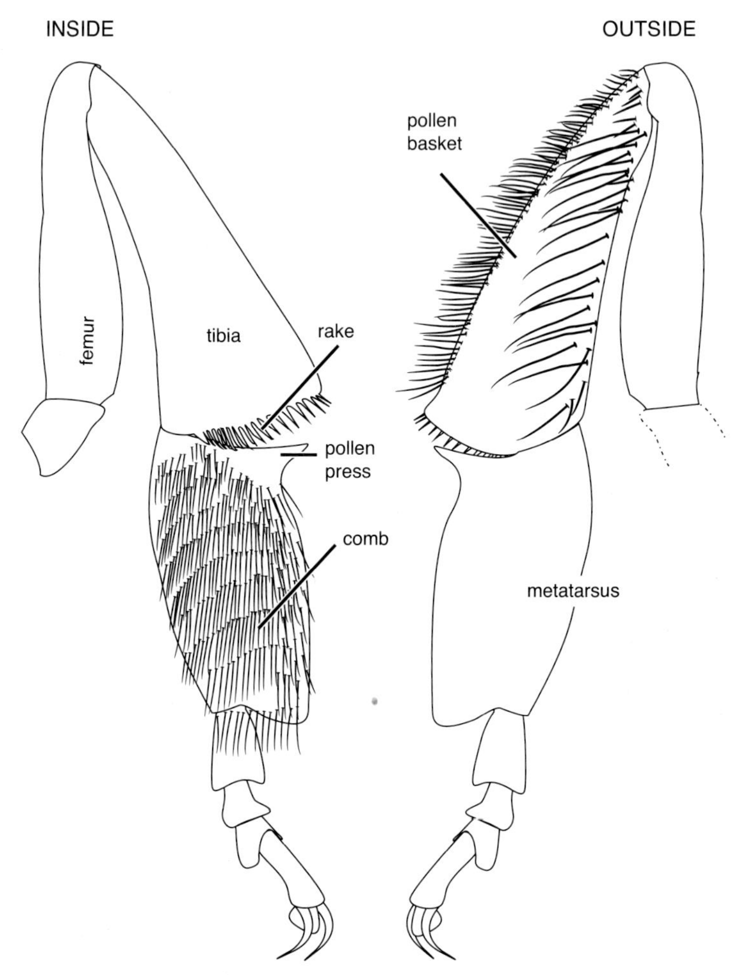

FIG. 18.3. Anatomy of the inside right hind leg and of the outside left hind leg of a worker honey bee.

other hind leg and then she pushes the rake downward. This compresses the pollen into a compact mass that sits atop a knobby structure, called the "pollen press," at the top of the metatarsus of the hind leg that the bee has just pushed downward. Once a suitable mass of pollen has been raked onto the pollen press on one of her hind legs, the bee closes the joint between the tibia and the metatarsus of this leg. This squeezes the pollen upward and outward. It passes to a concave surface called the corbicula ("pollen basket") on the tibia of the hind leg. Here the pollen mass is held in place by sturdy, curved hairs that ring the pollen basket.

Scraping together the sticky pollen and loading it into the pollen baskets is an intricate behavior. It is also a conspicuous behavior, because when a bee packs pollen into her pollen baskets she is apt to hover in place over the flowers she is visiting. Watching pollen collectors in action will give you a good view of a worker bee "showing off" her talent for producing rapid and precisely coordinated movements of all six legs simultaneously. Having just two legs to manage, I am humbled when I watch a worker bee packing her pollen baskets.

A worker bee is certainly adept at autogrooming, but there are some surfaces on her body that she cannot clean thoroughly on her own. These include the posterior (rear) surfaces of her head and the dorsal (top) surfaces of her thorax, especially the "hinges" at the bases of her wings. Bits of debris can get lodged in these hard-to-groom spots by accident, as when a bee is inside the hive and bits of crud fall onto her. The same thing can happen to foragers when they are outside the hive. Indeed, some plants, such as Himalayan balsam (*Impatiens glandulifera*), have flower structures that deposit pollen in the hard-to-groom spots by "design" (Fig. 18.4); doing so makes the bees more effective pollinators for these plants. So, there are times when worker bees sense that they need help with their grooming. They will then produce a strange-looking call for assistance: the grooming invitation dance.

To the best of my knowledge, the first detailed description of the grooming invitation dance was written by Professor Mykola H. Haydak. He was a Ukrainian scientist who came to the U.S. in 1930, earned his Ph.D. at the University of Wisconsin in 1933, and then worked as a professor of entomology at the University of Minnesota until 1966. His areas of expertise were bee nutrition and beekeeping. In the September 1945 issue of the *American Bee Journal*, he published a lovely article titled "The Language of the Honeybees." It includes vivid descriptions of the

FIG. 18.4. Honey bee collecting pollen from a flower of Himalayan balsam in Scotland. The anthers in this species' flowers deposit their pollen deftly atop a worker bee's thorax, where she cannot groom it off. The light stripe atop her thorax shows this limitation.

grooming invitation dance of a worker bee seeking assistance, and of the responses to it by fellow worker bees:

> Occasionally a bee feels the need of being cleaned. In such cases it performs a special dance, consisting of a rapid stamping of the legs and a rhythmic swinging of the body to the sides. At the same time the bee rapidly raises and lowers the body and tries to clean around the bases of the wings with the middle pair of the legs. Such

> a "shaking" dance may be observed any time during the year, even during the winter. Usually the bee which is closest to the dancer touches the latter with its antennae and begins to clean the dancer. With the mandibles widely spread the "cleaner" touches the thorax of the dancer just under the bases of the wings. As soon as the dancer feels the touch of the cleaner, it stops dancing, slowly spreads out the wings of one side, bends the abdomen and curves the body to the side and somewhat upward as if accommodating the cleaner. The latter works very energetically. Its antennae are held close to the mandibles. With shearlike motions of the mandibles it cleans around the base of the wings. From time to time it stops, standing on the last two pairs of legs, the front pair being held in the air, and works with the mandibles as if chewing something that it found while cleaning, holding the antennae close to the tips of the mandibles. Then the cleaner continues her work, "clipping" with the mandibles over the scutum from the rear to the front, sometimes over the head, and in the grooves of the thorax; sometimes she climbs on the dancer, crawls to the other side and cleans under the opposite pair of wings and then quits. If the dancer is satisfied with the work she cleans her tongue, antennae and the body in general and goes about her duties. In case the cleaning was not well done, the bee continues to dance, and either the same cleaner or another bee starts the process of cleaning all over again.

Haydak was not able to film the grooming invitation dance, so he was unable to make a precise description of the movements making up this behavior. Also, he did not investigate what causes a worker to perform this dance.

For 50 years after Haydak's paper, no further work was reported on the grooming invitation dance. Finally, in 1995, two biologists at the University of Ljubljana (in Slovenia), Janko Božič and Tine Valentinčič, reported a little study they had made of this dance using video analysis. They reported two quantitative features of this eye-catching behavior:

the median duration of a dance is rather short (just 8 seconds), and the probability (P) that a bee that has performed a grooming invitation dance will be groomed almost immediately by a nest mate is quite high ($P = 0.72$). The latter finding strongly supports the notion that this behavior is a signal whereby a bee solicits grooming assistance.

The most recent investigation of the form and function of the grooming invitation dance was made by an undergraduate student at Cornell University, Benjamin B. Land, who examined it further in 2002. Ben had helped me with my summer research projects that year, and in late August he asked to pursue a project of his own, so he could fulfill the requirements for an honors degree in biology. He was on track to graduate in May 2003, so he needed a research project for which he could collect all the data in September and early October. I encouraged him to make a fine-grained study of the movements of a worker bee's body when she produces the grooming invitation dance. This would involve making close-up, digital video recordings of bees producing this dance and then patiently analyzing the recordings using slow-motion playback. I also encouraged Ben to test experimentally the hypothesis that debris lodged in a worker's wing "hinges" stimulates her to perform the grooming invitation dance.

To perform this experiment, Ben worked with a small colony of honey bees living in an observation hive, from which he gently removed the glass sheet covering one of its sides. Next, he selected at random, *and one at a time*, 60 worker bees living in this hive. He did not handle any of these bees. But he did puff a tiny amount (0.5 milligrams) of chalk dust from the tip of a glass pipette onto the thorax of each of 30 bees, one by one, at the bases of her wings, and then video recorded her behavior for 10 minutes. This experiment needed a control, so for every experimental bee that Ben puffed with chalk dust and watched, he watched another bee that had received just a puff of air with no chalk dust. Ben then watched each bee for 10 minutes after he puffed her (with chalk dust or clean air) to see whether she groomed herself, or performed the grooming invitation dance, or did both. Ben managed to follow each control bee for

FIG. 18.5. *Left*: The grooming invitation dance. *Right*: The allogrooming response that it elicits. The grooming by another bee almost always involves the groomer mounting the groomee as soon as she has stopped dancing and has spread one or both wings. Note: when a bee grooms another bee, the groomer tilts her head so her mandibles point forward instead of downward. This is why the heads of the groomed bee and the groomer bee look so different in the right-hand part of this figure.

10 minutes by never taking his eyes off her after puffing her with clean air. Impressive!

Ben analyzed 25 grooming invitation dances. This work yielded several solid findings. *First*, it revealed that when a bee produces this dance, she does not walk across the comb like a waggle dancer or a tremble dancer. Instead, she stands rooted in one place with her six legs widely spread and with each leg's tarsal claws gripping the comb. This stance makes sense; it prevents her from falling off the comb as she heaves her body mightily and rapidly (4 times per second) from side to side on the vertical surface of a comb (Fig. 18.5). *Second*, it confirmed that grooming invitation dances are short-lived. Their average duration is just 9 seconds. In comparison, waggle dances have an average duration of one minute and tremble dances have an average duration of 27 minutes.

Third, it showed that when a bee performs a grooming invitation dance, she may pause briefly to groom herself. Sometimes she grooms her abdomen, by raising it and pivoting it sideways, and then brushing it with her hind legs. Other times she grooms her head or thorax with her forelegs and middle legs. I suspect that when a bee makes a pause in her grooming invitation dance to groom herself, she is trying to remove something especially troublesome from her body. She can't wait for help!

Ben's analyses of his video recordings also documented the effects of the grooming invitation dance on nearby bees. When Ben watched 65 bees that *had performed* this dance, he saw that 45 of them (69 percent) received allogrooming by adjacent bees within 30 seconds of the start of their dancing. (Note: this percentage nearly matches what Janko Božič and Tine Valentinčič reported in their 1995 paper: 72 percent). But when Ben watched (for 30 seconds each) 65 bees that *had not performed* a grooming invitation dance, he never saw them receive allogrooming. I like to say that results like these—45 of 65 vs. 0 of 65—pass the "interocular test" for statistical significance; their significance strikes you between the eyes! Indeed, the probability of getting such a stark difference by chance is less than 0.00001. There can be no doubt that when a bee performs the grooming invitation dance, she is issuing a request to be groomed.

The results of Ben's particle-puffing experiment were also crystal clear. Of the 30 bees that received a puff of chalk dust, 28 groomed themselves and 15 performed a grooming invitation dance. Of the 30 bees that received a puff of clean air, only 6 groomed themselves and just one performed a grooming invitation dance. These results strongly support the idea that debris on a bee's body can induce a bee to perform the grooming invitation dance.

One question about this dance still lacks an answer: How does a worker bee, standing in the darkness of a colony's nest, sense the grooming invitation dance of a nearby nest mate? By sensing air movements produced by the thrashing movements of the dancer's body? By touching the dancing bee and so sensing directly her back-and-forth movements? Or perhaps by sensing substrate (comb) vibrations? When a bee performs the grooming invitation dance, she must load into the comb much of the energy of her body's side-to-side thrashing, so it seems likely that she produces a strong comb-vibration signal. And because the body thrashing of the grooming invitation dance has a markedly lower frequency (only about 4 Hz [cycles per second]) than the body waggling of the waggle dance (about 15 Hz), the comb vibrations of these two kinds of dance should be easy for the bees to distinguish. So I think it is likely that low-frequency comb vibrations are the primary channel of communication between the bees sending and receiving the grooming invitation signal.

I close this chapter with a speculation on the evolutionary origins of the grooming invitation dance. Because the form of this dance resembles the form of a worker bee's behavior when she grooms herself (i.e., performs autogrooming), I suspect that the precursor, or ancestral form, of the grooming invitation dance is the set of movements made by worker bees when they perform autogrooming. These movements are a reliable indicator (a cue) that the bee making them senses that she needs grooming. Then, at some point in the past, worker bees probably evolved an allogrooming response to fellow workers that were grooming themselves especially strongly, i.e., to bees that were having great difficulty getting themselves clean. Since workers living together are nest mates, both the bees receiving the grooming and the bees giving the grooming would benefit from this social response. Thus the stage was set for natural selection to favor refinements both in signal production (exaggeration of the self-grooming behavior) and in signal reception (greater sensitivity to bees producing the self-grooming behavior). The eventual result is, I believe, the remarkable grooming invitation dance and the reliable allogrooming response that are fun to watch today.

CHAPTER 19

Colony Thirst

One of the joys of studying honey bee behavior is to watch a worker bee who is doing something that seems strange, and then to discover (eventually) that what she was doing actually makes perfect sense. Her behavior was contributing to the smooth functioning of her colony. So, I treasure the times when I spot a bee acting in a way that puzzles me. I look at her curious behavior as a signpost that is pointing to a good mystery. In Chapters 15 and 17, we saw how this happened when I saw bees producing tremble dances and shaking signals. In this chapter, we will look at a third example of how watching bees doing something puzzling led me to pursue some detective work. The mystery here was how a honey bee colony regulates its water intake.

My curiosity about this part of a colony's functioning was aroused back in May 1985. I was devoting several days to studying the crop ("honey stomach") contents of worker bees that were landing at their hive's entrance with their abdomens bulging. I thought that all these bees were nectar foragers, and I was examining their crop contents because I wanted good information about the range of sugar concentrations of the nectar loads brought home by these bees. I was about to start performing experiments in which I would use sugar-water feeders to give a colony "nectar" sources with different levels of profitability, and I wanted the sucrose solutions in my feeders to have realistic concentrations. These experiments would begin my analysis of how a colony's nectar foragers wisely allocate themselves among the flower patches dotting the landscape around their home (discussed in Chapter 12).

My method for collecting data on the crop contents of returning bees was simple. I sat beside a hive housing a strong colony of bees, plucked one bloated forager at a time from the hive's landing board, and then gently squeezed her swollen abdomen so she would regurgitate a droplet onto a hand-held refractometer that was calibrated for sucrose solutions. After releasing the bee so she could scurry inside the hive, I read the refractometer to learn the sugar concentration of her nectar load, and I entered this datum in my notebook. Then I plucked another bloated bee running into the hive . . .

I made 681 measurements over a string of five days of warm, sunny weather in the middle of May 1985. Figure 19.1 shows what I found on these five days, plus three more days in late June 1989. The bees' nectar loads had a wide range of sugar contents: from 15% to 65% sugar by weight. (A chemist would say that these nectar loads were like sucrose solutions that ranged in concentration from 0.5 to 2.5 molar.) This graph showed me the range of concentrations of sucrose solutions that I could put in my sugar-water feeders without giving the bees something abnormal.

This graph also showed me something that seemed odd. There were 44 bees (about 5 percent of the total sample) that had brought home "nectar" loads with remarkably low sugar contents, just 0–2%. Basically, these were loads of water. This seemed odd because I made the first set of 681 readings on warm, sunny days in the middle of a heavy nectar flow from dandelions (*Taraxacum officinale*), so I figured the colony was getting plenty of water from dandelion nectar on these days. Moreover, I made the second set of 154 readings during a string of three sunny days when the colony was collecting nectar from sumac trees (*Rhus glabra*). Both times, I wondered: Why did some bees come back with their crops filled with water? When nectar is plentiful, doesn't a colony get all the water it needs from the loads of nectar that its foragers bring home? The data showed clearly that the answer to the last question is "No."

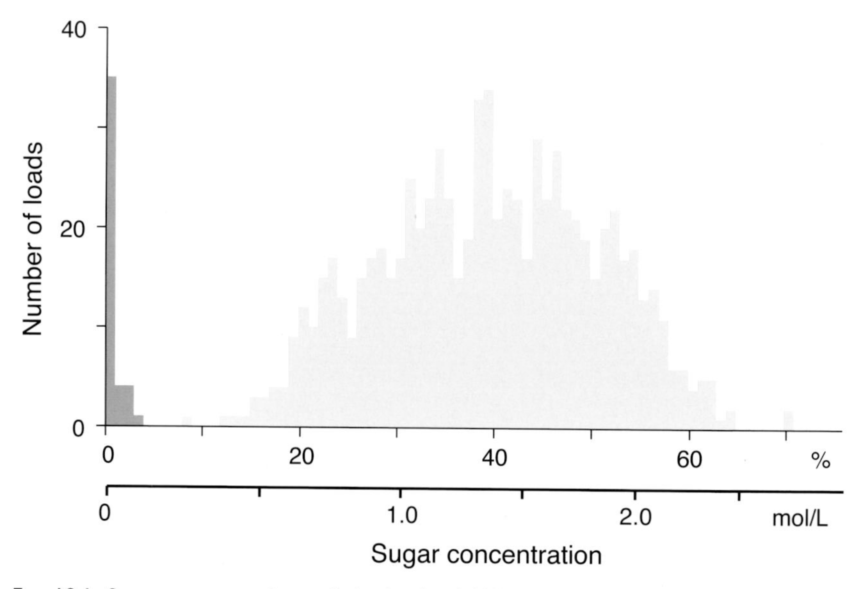

FIG. 19.1. Sugar concentrations of the loads of 835 bees that landed at their hive's entrance with noticeably swollen abdomens. Blue bars: water. Yellow bars: nectar. Collection dates: 12–16 May 1985 and 28–30 June 1989.

Pondering this graph was my first step toward understanding how the water economy of a honey bee colony works.

What I began to realize is that a colony's nectar-collection and water-collection operations work in parallel and somewhat independently. A high intake of nectar does not fully meet a colony's need for water. Also, I started to see that this makes perfect functional sense, because these two liquids serve different purposes. Nectar provides energy. Water provides hydration. Collecting nectar lowers *colony hunger*. Fetching water lowers *colony thirst*. Furthermore, these two liquids have different supply-demand relationships. With nectar, the supply is highly variable (nectar flows start and stop), and the demand is nearly constant (almost always, a colony tries to amass more honey). With water, however, the supply is stable (springs and other wet spots offer plentiful water at all times), and the demand is variable (a colony experiences hot days and cool days).

While thinking about how a honey bee colony organizes its collection of nectar and water, I recalled that one of my mentors, Martin Lindauer, had described in his wonderful book *Communication among Social Bees*, that for water, as for nectar, there is a division of labor between the bees that work outside the hive doing the *collecting* and those that work inside the hive doing the *consuming*. Lindauer knew about this because in the winter of 1951 he had moved a small honey bee colony living in an observation hive into a greenhouse at the University of Munich. He had also set up in this greenhouse a water source for his colony: a tray holding a soggy clump of moss. So, by measuring the rate at which the colony's water collectors made trips to the moss, he could measure the colony's rate of water collection. He then imposed a heat stress on the colony. He did so by shining the light from an incandescent lamp onto the combs in the observation hive. The colony's response is graphed in Figure 19.2. During the hour before the heat stress, the air inside the observation hive was cool, about 68°F (20°C), and only two bees visited the drinking place. But after the lamp was turned on, the air temperature inside the hive rose to 100°F (38°C), and soon the colony's water collectors sprang into action.

Lindauer reported that when the hive became overheated, he observed the following whenever a water collector dashed inside the hive: three or four bees would rush up to her, eagerly suck up the water droplet she extruded between her mouthparts, and then use the water to cool their home. They did so partly by smearing water on the combs for evaporative cooling, and partly by gobbeting. This is the behavior depicted in Figure 19.3. A worker bee extrudes a water droplet from her mouth and then she draws it out with her proboscis to make a thin film, which has a large surface for evaporative cooling. (I believe that Othniel W. Park, an entomologist at the University of Illinois, was the first person to describe this behavior in detail, in 1925.) Lindauer also reported what he observed after he turned off the lamp. The hive temperature dropped, the water collectors were no longer begged stormily, and within a half hour these bees had stopped visiting the water source.

I let these thoughts about water collection by honey bees lie dormant until the spring of 1993. This is when a student from the University of

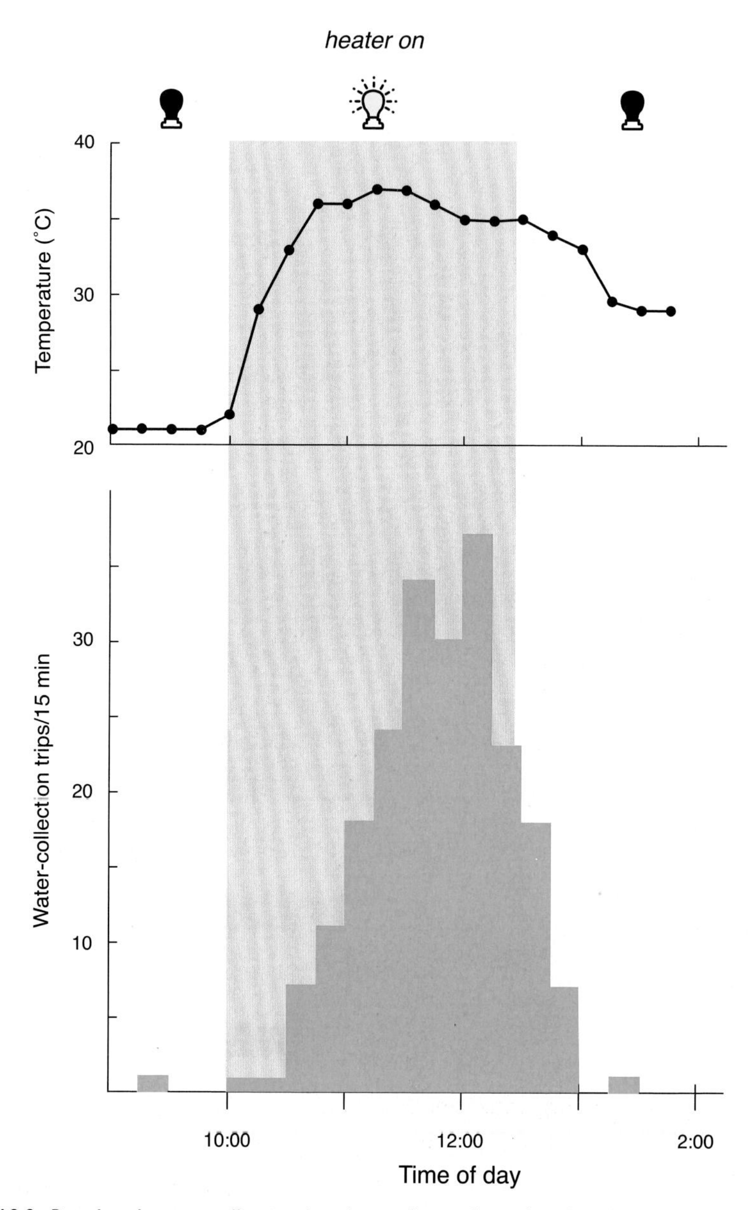

FIG. 19.2. Regulated water collection by a honey bee colony. A colony living in an observation hive was moved into a greenhouse, where it was possible to monitor closely the traffic level of its water collectors. A clump of wet moss provided a water source. When an incandescent lamp beside the hive was turned on, the temperature inside the hive rose to 100°F (38°C), and the traffic of water collectors rose rapidly to support the colony's evaporative cooling of its home. This lowered its temperature. When the lamp was turned off and the temperature inside the hive dropped, the water collectors became less active.

FIG. 19.3. Gobbeting by a middle-age worker bee functioning as a water handler. She has spread her mandibles and has extended her proboscis, but she has not extended her tongue. Instead, she has regurgitated a droplet of water and, to draw this water into a film, she has swung outward the proboscis structures (the galeae) that form a sheath around the tongue when it is extended. This creates a large surface for evaporation and cooling.

Würzburg in Germany, Susanne Kühnholz, joined my laboratory group to conduct a research project. Back then, students at German universities needed to complete four years of coursework *and* produce a research dissertation to earn the university degree called the Diplom. Susanne had helped Professor Martin Lindauer—then an emeritus professor at the university—with a behavioral study of honey bees, had enjoyed this work, and wished to do her *Diplomarbeit* (equivalent to a Master's thesis) on honey bee behavior. Also, my Ph.D. thesis advisor, Professor

Bert Hölldobler, had moved recently from Harvard University to the University of Würzburg. So, I was happy to host Susanne at Cornell. We considered several topics, and found that we were both keen to further investigate the control of water collection in honey bee colonies.

Our investigation would build on the pioneering work on this subject that Martin Lindauer had done in the early 1950s at the University of Munich. Besides describing the water-collection responses of whole colonies (e.g., Fig. 19.2), Lindauer had studied the behavior of individual water-collector bees. He reported that once a water collector has started her work, she gets an update about her colony's water need each time she delivers a load of water, and she responds to this update accordingly. If the water need remains high, then she continues collecting and may even perform waggle dances to recruit others to the task. If the water need has dropped, then she ceases collecting. How, exactly, a water collector senses whether her colony's need for water remains high or has dropped remained a mystery.

Susanne and I decided that she should focus on investigating how a water collector knows when to *stop* fetching water. Lindauer had noticed, while performing the experiment whose results are shown in Figure 19.2, that when the heater lamp was turned off, the receiver bees showed less interest in the water loads offered by the collector bees. In Lindauer's words, water collectors "now have to run around in the hive themselves, trying to find somewhere a bee that will relieve them of at least part of the water load." Susanne and I set ourselves the goal of documenting in detail the changes that a water-collector bee experiences when her colony's need for water drops from high to low. Lindauer had reported one change: a marked increase in the amount of time that passes between when a water collector enters the hive and when her honey stomach (crop) becomes empty. He called this the bee's *Abgabezeit*, which translates to "giving-off time" or "delivery time."

Lindauer's study showed that a water collector might pay attention to her delivery time to sense her colony's need for more water. But his study also showed that she might pay attention to some other variable. Here are the other variables that Susanne and I suspected might be important to a water collector: *One*, how long she has to search to find a bee who will start to take her water. *Two*, how many bees unload her simultaneously (it ranges between one and four). *Three*, the liveliness of the stroking of her antenna during the unloading process. *And four*, how many unloading rejections she experiences before she finds a willing receiver. A water collector experiences an "unloading rejection" when she is met by another bee who comes up, unfolds her tongue, and probes the droplet that the water collector has extruded between her mandibles, but then quickly withdraws her tongue as if to say "Pardon me, but I am seeking nectar, not water." It was possible, too, that a water collector pays attention to some combination of these four variables and the one highlighted by Lindauer: delivery time.

To see which possibility is the reality, we needed to measure all these variables of the unloading experiences of water collectors, and we needed to do so both when a colony's need for water is high and when it is low. We hoped that one of these five variables would stand out as the clearest, and therefore probably the most important, indicator to a worker bee of her colony's need for more water.

Susanne and I knew what we needed to do to get started. We must move a colony into an observation hive, then heat the hive with a lamp to stimulate some of the colony's members to collect water, and then label these water collectors so we could study their behavior. If we could do all these things, then we would be able to track individual water collectors over time and see how their collecting activities and their unloading experiences changed when we shifted their colony's water need from high (heating lamp on) to low (heating lamp off).

So, Susanne and I installed an observation hive that held a colony of some 3,500 bees in the Dyce Laboratory for Honey Bee Studies, which is just a 10-minute walk north of my laboratory in the Liddell Field Station. Then we heated Susanne's observation hive with an incandescent lamp. Soon we saw bees dancing wildly inside the hive to advertise a boggy place that is less than a quarter of a mile (400 meters) south of

FIG. 19.4. Tim Judd, putting paint marks on water collectors when they came to the boggy spot that was a popular watering hole for the honey bee colonies at the Dyce Laboratory and the Liddell Field Station. Both laboratories are only a few hundred yards (meters) away.

Dyce Lab, and is near my laboratory. But now we faced the difficulty of getting a label on even one of the water collectors from the small colony living in Susanne's observation hive. There were several full-size colonies at the Dyce Lab, and about a dozen more at my laboratory, so we figured that almost every water collector that visited the bog came from one of the full-size colonies at the two bee labs. Nevertheless, Susanne persisted. Eventually, and with the help of the Cornell student Tim Judd, who was fearless about wading barefoot into the bog to label many dozens of water collectors (Fig. 19.4), Susanne got one water collector from her colony labeled. This bee carried a splotch of yellow paint on her thorax, so we named her Yellow Thorax. I have difficulty remembering names of people, but I do recall the name of this bee.

On 29 July 1993, Susanne performed the experiment whose results are shown in Figure 19.5. It shows the changes in collecting activities and unloading experiences of Yellow Thorax when her colony's need for water

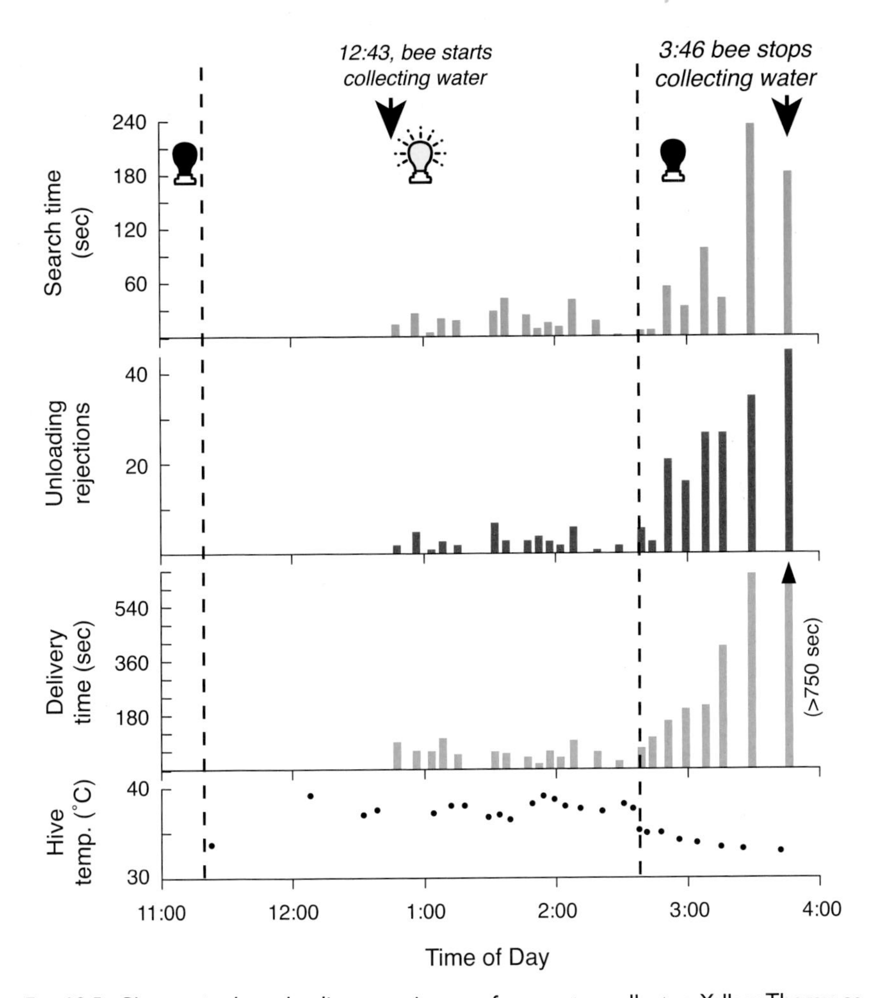

FIG. 19.5. Changes in the unloading experiences of one water collector, Yellow Thorax, as her colony's need for water went from high to low. When the observation hive was heated and the colony had a high need for water to prevent its brood nest from overheating, she experienced three things: (1) short search times to find a bee eager to take her load of water, (2) few encounters with bees that rejected her load of water, and (3) short delivery times. But when the heating lamp was turned off at 2:38 (dashed line followed by black bulb at right), so the need to cool the brood nest declined, the water collector experienced longer search times (green bars), more unloading rejections (red bars), and longer delivery times (blue bars). These findings show that there are several features of a water collector's unloading experience that can inform her of her colony's need for more water.

rose, then stayed high for more than two hours, and eventually subsided. At 11:20 a.m. the heating lamp was turned on, and by 12:10 p.m. the temperature inside the observation hive was dangerously high, nearly 104°F (40°C)! At 12:43 p.m., Yellow Thorax made her first excursion for water. At 2:38 p.m., the heating lamp was turned off and the hive's interior temperature quickly fell to normal, but Yellow Thorax continued to fetch water. We see that she made 14 collecting trips while the heating lamp was on, and then 8 more after the lamp was switched off. We also see that so long as the hive was being heated, Yellow Thorax searched only briefly (average, 21 seconds) to find a hive mate who wanted some of her water; she encountered only a few bees (average, 3 individuals) who rejected her offer of water; and she needed only a minute or two total to deliver her load of water. We see, too, that once the danger of nest overheating had passed, her search times, unloading rejections, and delivery times all increased markedly. She stopped her work after she walked into the hive with a load of water around 3:45 p.m., then spent 10 minutes offering water to more than 40 hive mates, but experienced only rejections. Finally, she walked to the edge of one of the combs and then stood there almost motionless. Given what we now know about sleep in honey bees (see Chapter 17), I believe that she was tired from hauling home 22 loads of water, each one nearly as heavy as herself, and had decided to retire to a quiet corner of the hive to take a nap.

Thinking back about Yellow Thorax leads me to wonder about the level of awareness that this worker bee had about the things that happened in her life on this day. I doubt that she experienced *thoughts* about the danger of her colony overheating, but I think it is possible that she experienced *feelings* about her colony's urgent need for water. At first, she might have felt too hot. Then, further along, she might have felt very thirsty. And when the danger had passed, I think it is possible that she felt exhausted, and maybe even sleepy. I recognize, of course, that Yellow Thorax was a honey bee, not a human being. But the way that she slowed her movements and withdrew to special place, a quiet corner of the hive, when the emergency was over makes me suspect that she was feeling a need for rest.

The experiment conducted in 1993 showed us that when a worker bee comes home with a load of water, she has several possible ways of sensing her colony's need for more water. She might note how much time passes between when she enters the hive and when she finds a hive mate that will take her water load (*search time*). Or she might note how much time passes between when she enters the hive and when she finishes her unloading (*delivery time, or* what Lindauer called the *Abgabezeit*). A third possibility is that she notes how many times she experiences an *unloading rejection*. This is when another bee inserts her tongue between her (the water collector's) mandibles but then quickly withdraws her tongue and walks away. This might happen because the other bee is seeking sweet nectar, not bland water. A fourth possibility is that she senses the *maximum number of simultaneous unloaders*, i.e., the largest number of unloaders whose tongues are inserted simultaneously between her mouthparts. It is possible, too, that a water collector registers more than one of these variables, and forms a gestalt (integrated) sense of her colony's need for water.

To find out which variable changes most strongly, and so perhaps gives a water collector the clearest indication of her colony's need for water, we repeated the experiment shown in Figure 19.5 several more times the following summer. This time, we worked at the Cranberry Lake Biological Station (CLBS) instead of the Dyce Laboratory. Changing our study site made it much easier to perform our experiment because at the CLBS there was just one honey bee colony for miles and miles in all directions . . . our study colony! So, whenever we heated the observation hive and then spotted a worker bee licking water from a pebble along the lake shore or a tuft of wet moss (Fig. 19.6), we knew she came from our study colony. This was a game changer for us. Now, it was possible to perform seven more trials of the experiment that we had struggled to pull off just once the summer before.

FIG. 19.6. Water collector imbibing water from the stems and leaves of a moss plant.

It was exciting to assemble the data for all eight trials of our experiment, and then calculate how strongly, on average, each feature of the unloading experiences of our eight focal water collectors changed when we shut off the heating lamp and lowered the colony's need for water. We learned four things: (1) the maximum number of simultaneous unloaders did not change, (2) the mean delivery time increased by a factor of 4, (3) the mean search time increased by a factor of 6, and (4) the mean number of unloading rejections increased by a factor of 10! So, it appears that the number of unloading rejections is the strongest indicator to a water collector of her colony's level of need for more water. This indicator is based on something conspicuous—another bee breaks contact and walks away—and it has a high range of values. For example, one of the eight water collectors that we watched experienced only 0.9 rejections (on average) upon return to the hive when the colony's water need was high (heating lamp on), but then she experienced 13.7 rejections (on average) when its water need was low (heating lamp off). We must keep in mind, though, that a worker honey bee is capable of integrating multiple pieces of information about an experience to make an overall evaluation of it. You may recall from Chapter 3 that this happens when a nest-site scout conducts an inspection of a potential homesite. So, I think it is possible that a water collector registers several variables of her unloading experience and integrates this information to get an accurate sense of her colony's need for water.

So far, we have focused on the mystery of how a water collector knows when to *stop* her work. Let's turn now to the puzzle of how a water collector knows when to *start* her work. We can safely assume that her decision to start collecting water comes from feeling thirsty, that is, from feeling an urge to ingest water. It is likely that this urge to drink water is stimulated by sensors in a bee's brain that have registered an increase in the osmotic pressure of her blood (the technical term for an insect's blood is "hemolymph"). If so, then a bee's sense of thirst arises by a physiological mechanism that is similar to what underlies our own sense of thirst: osmoreceptors in the brain register a rise in the concentration of solutes in the blood, mainly sodium chloride and sugars. (Note: the walls of the honey stomach [crop] in a worker bee are impermeable to water. This is critically important; it prevents a worker bee's blood [hemolymph] from becoming diluted when she loads up with water or thin nectar. In other words, it protects her from osmotic shock.)

My first experience with extremely thirsty worker bees occurred on a morning in January 1995, when I came to my office in the Liddell Field Station. Deep snow covered the land, but on that morning strong sunlight had warmed the air, and some of the bees living in the observation hive that I had in my office were scurrying down its entrance tunnel and flying off. I heard their activity and supposed that they were heading out to make "cleansing flights" (defecation excursions). But when I removed one of the hive's cover-boards, I saw several bees feverishly performing waggle dances, all of which had very short waggle runs. This told me that these dances indicated a site just outside the building. For a moment I was puzzled, but soon I understood. These bees were water collectors! They were advertising puddles of snowmelt on the dark asphalt of the parking lot. I saw, too, that whenever one of these water collectors dashed into the hive, she was mobbed immediately by fellow workers that pressed around her and poked their tongues between her mandibles. Each one was straining to get a sip of water. I saw further that when these "celebrity" water collectors finished unloading, they performed marathon-class waggle dances. Indeed, one of these bees produced the longest-lasting waggle dance I have seen. She danced for 339-plus dance circuits, thus for more than 5 minutes. I have to report the circuit count of her dance as "339-plus" because I did not see this bee start her dance. How long did it run, start to finish? Perhaps 500 circuits? Maybe even more?

This midwinter experience of watching desperately thirsty bees taught me an important lesson: when I keep a colony in a warm room over winter, I must squirt water into its hive from time to time, to replace the water the colony would normally get from condensation on the walls and combs inside its hive. This experience also made me keenly aware of how colonies living in cold-climate regions must benefit from having this "indoor plumbing." The condensation of water inside their hives helps these colonies meet their need for water over winter without their water collectors undertaking risky flights to outdoor water sources when it is dangerously cold outside.

There is a valuable lesson here if you are a beekeeper and you live where it is risky for your bees to fly from their hives in winter: *Do not provide openings near the tops of your hives to reduce water condensation inside them.* Colonies benefit from this condensation as a handy, indoor source of clean water over winter. Colonies benefit, too, from not having lots of heat leak out the tops of their homes when it is cold outside. I am sure that the benefits of having clean water handy and living in a tight home help explain why honey bees work hard to fill with propolis any cracks in the walls and ceilings of their hives (to be discussed in the next chapter).

On the hot days of summer, though, a colony's water collectors certainly do need to spring into action outside the hive, for this is when their homes can be threatened with overheating. In 2015, two of my students—Madeleine Ostwald and Michael Smith—and I looked closely at how water collectors are roused when their colony's brood nest is in danger of getting too hot. We knew that to investigate the activation of a colony's water collectors, we needed three things: (1) control of a colony's access to water, (2) a way to measure precisely a colony's rate of water collection, and (3) the means to observe closely the actions of the colony's water collectors inside their home. So, we worked with a small colony living in an observation hive that we had moved into a room in a greenhouse (Fig. 19.7). We provided this colony with a reliable and accessible water source: one of my sugar-water feeders (see Fig. 13.4) that we filled with water. By putting this water source on a sensitive balance-beam scale, and making readings of its weight every 30 minutes, we were able measure precisely the colony's rate of water collection.

We stimulated the colony's water collection by raising the temperature inside its hive, which we did by shining the light from an incandescent lamp onto one side of its lower comb. This comb held most of the colony's brood. Then, when its water collectors began visiting our water source, we labeled them with paint marks. We needed to recognize each one as an individual. The plan for our study was simple: observe the in-hive behavior of the colony's water collectors across a long (6-hour) block of daytime. In each run of the experiment, the 6-hour observation time would consist of an initial 2-hour period without heat stress (lamp off), then 2 hours with severe heat stress (lamp on), and finally 2 hours of recovery (lamp off). This procedure enabled us to track how the water collectors in our colony sprang into action, then worked furiously to fight the danger of their brood nest overheating, and finally stepped down when the danger had passed.

FIG. 19.7. Madeleine Ostwald taking data for our study of what stimulates water collectors to action when their colony's nest is threatened by overheating. The colony was being heated with an incandescent lamp, its water collectors were visiting the water source on the balance-beam scale, and Madeleine was tracking labeled water collectors (one at a time) inside the observation hive. Some bees had evacuated the hive, forming a "beard" of bees around the hive's entrance.

What did we learn? One finding was that the bees that functioned as the colony's water collectors did not start their work immediately after we began heating their home. Even when the temperature inside the hive rose rapidly to a dangerously high level (104°F/40°C), they did not visit the water source until an hour or so had passed. Clearly, it was not high temperatures inside their home, per se, that roused the water collectors to action.

So, what did spur these bees to action? There are several possibilities. One is that a water collector gets reactivated by an *internal physiologi-*

cal signal, such as feeling extremely thirsty. A second is that she gets reactivated by an *external social cue*, such as being begged for water. And a third is that she springs into action only when she senses both indicators of water need in her colony: she gets begged for water but she has none to share (and so feels thirsty).

We were not able to study the physiological signals of water need (personal thirst) inside our water collectors, but we were able to look closely for social cues that might reactivate these bees. We did so using a method called "making focal observations." This consisted of continuously watching one water collector at a time for 30 seconds and recording everything she did. (Each of our focal bees had been seen collecting water the day before, so she was already labeled for individual ID.) We recorded whenever the bee began doing any of the following nine behaviors: standing (still), walking, grooming self, grooming other, waggle dancing, begging for food/water, being begged for food/water, giving or receiving food/water, and inserting her head into a cell. We watched each water collector once or twice every 10 minutes during the 6 hours of observation in each experiment. (*Technical note: the observer usually wore a headset with binocular magnifying lenses. This was helpful, but even so this work was exhausting.*)

The work of watching the water collectors over and over during each 2-hour stage of the experiment showed us that after an hour or so of their home being heated by our lamp, the water collectors began to get approached by worker bees *that were not water collectors*. When the two bees came together, face to face, the unmarked bee (not a water collector) would extend her tongue and then touch the water collector's mouthparts. Sometimes, this was followed by the water collector regurgitating water and the two bees staying in contact for several seconds. But other times, the contact was brief, just a second or two, probably because the water collector had no water to share. A brief contact was not long enough for a substantial transfer of water, but probably it was long enough to inform the water collector that she had been begged for water.

We concluded that when a water collector has given away all the water in her crop, and then she receives still more requests, she is stimulated to collect another load of water.

Further studies are needed to know for sure whether a water collector is reactivated to water collection by just an internal indicator (e.g., feeling thirsty) or just an external indicator (e.g., being begged), or by their combination. I expect that we will learn that she has to experience both things to be stimulated to resume her work. In other words, I think it is only when both signs of water need are "flashing" in a worker bee's brain that she knows that her colony really, really needs more water.

CHAPTER 20

Resin Work

A honey bee colony builds its nest mostly of beeswax, a material that worker bees produce themselves, but a colony also uses a material that the workers collect from the environment: resins. These are sticky substances that are secreted from various places on plants, but especially from leaf buds on deciduous, broad-leaved trees, and from wounds on coniferous, cone-bearing trees. I have enjoyed watching bees collect light-yellow resin from the leaf buds of an eastern cottonwood tree (*Populus deltoides*) that grows near one of my apiaries. It is a handy source for my colonies there. I am not sure, but I suspect that worker bees find their resin sources by orienting upwind when they smell the scents of plant resins, many of which have the unmistakable aroma of turpentine (a distillate of pine resins). Plants secrete resins to defend themselves from herbivores, bacteria, and fungi, so it is not surprising that resins are laden with chemical compounds that are distasteful to herbivores and are toxic to bacteria and fungi.

Sometimes the words "resin" and "propolis" are used interchangeably, but this is a mistake. The word "resin," like the words "nectar" and "pollen," is a botanical term. The word "propolis" is a beekeeping term. It was coined by writers in Ancient Greece: *pro* (in front of, or at the entrance to) and *polis* (city or community). Honey bees make propolis by mixing resin with beeswax; its composition is 50–80% resin, and 20–50% beeswax. I suppose the bees do this mixing because it saves on resin collection. Perhaps, too, the bees find it easier to "handle" propolis than straight resin.

Colonies will varnish smooth surfaces inside their homes with thin coatings of resin, but wherever there are cracks, rough surfaces, or needless openings, the bees will fill them with seams, coatings, or walls of propolis. Why do they do so? One reason is to improve colony health. Studies by Professor Marla Spivak and her colleagues at the University of Minnesota have found that colonies living in hives whose inner walls are coated with propolis, relative to colonies living in hives without much propolis on their walls, have healthier workers and better survival.

A second reason for this propolis work is to make the nest cavity more snug by making it less drafty. Where I live, in upstate New York, I see bees collecting resin and working propolis mostly in late summer and early fall, thus in August, September, and October. This is when my bees are closing cracks in my hives, sealing their covers tight, and erecting walls to reduce their entrances (Fig. 20.1). The bees are tightening up their dwellings in preparation for the cold and windy days ahead. I admire honey bee colonies for many things, but especially for their foresight in getting ready for winter far in advance of its arrival. I try to do the same with respect to filling my woodshed.

It is remarkable that worker bees can chew, handle, and carry bits of resin (and propolis) without it getting stuck to their mouthparts, legs, and other parts of their bodies. How is this possible? We know, from chemists' studies of adhesives, that the strength of adhesion between two materials depends on the molecules on their surfaces. If their surface molecules are dissimilar, then the molecular force of attraction between the two materials is low, so there is poor adhesion. This is why water puddles up on the freshly waxed hood of a car. Water molecules are highly polar, and wax molecules are apolar, so water and wax repel each other. I know nothing about the surface chemistry of the mouthparts and legs of bees, but I presume that it is very different from the chemistries of the resins they collect and the beeswax they secrete. Neither resin nor beeswax sticks to the bees' mouthparts, and neither does the resin-beeswax mix that we call propolis.

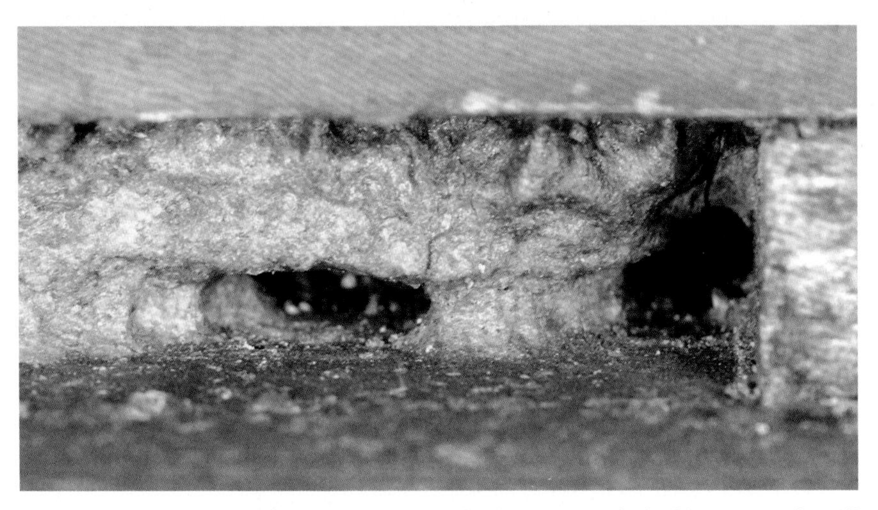

FIG. 20.1. A hive entrance that the bees reduced in late summer by building a propolis wall across most of its opening. They left just two small passageways; each provided space for just two or three bees to squeeze through simultaneously. Before the propolis wall was built, the dimensions of this hive's entrance were 0.75 × 2.25 inches (1.9 × 5.7 centimeters), and its area was 22 times larger than the total area of the two openings shown.

The focus of this chapter is the behavior of the worker bees that function as a colony's resin collectors. We will examine who these bees are, how they behave, and how they regulate their work. Little was known about the last topic until it was investigated in the early 2000s by Dr. Jun Nakamura, a professor at the Honeybee Science Research Center of Tamagawa University, in Japan. Jun and I met in the spring of 2000, when I traveled to Japan to present a set of lectures at his university. He and I quickly became friends. Then, in 2002, Jun came to Cornell to investigate the control of resin collection. I had studied how a colony controls its collection of nectar and of water, and I had observed (incidentally) bees coming into my observation hives bearing shiny loads of resin (Fig. 20.2). But I knew nothing about how a colony controls its collection of resin. So I was delighted when Jun decided to join me for a year and investigate how the resin sector of a colony's economy works.

One thing about resin collection that had been studied previously, and in detail, is how a worker bee packs the loads of resin onto her corbiculae (pollen baskets). Unlike the process of pollen packing, which is done

FIG. 20.2. Worker honey bee with a shiny load of resin on the corbicula (pollen basket) of her left hind leg. She has managed to bite off bits of resin, transfer them to her corbiculae, and bring home twin loads of this valuable substance.

while a worker bee hovers in midair, the process of resin packing is done while the bee stands on a surface. Often, this surface is her resin source. Dr. Waltraud Meyer, a biologist who studied honey bees at the Free University in Berlin in the early 1950s, examined the process of resin loading as part of her Ph.D. thesis on the *Kleinbauarbeiten* (small building projects) of honey bees. She described this process in terms of four stages: (1) using the mandibles to nibble off a resin chunk (on a cool day) or to pull off a resin strand (on a warm day); (2) using the forelegs to knead it into a lump; (3) transferring the lump from between the forelegs to the inner surface of the large, first-tarsal segment of one of her middle legs; and finally, (4) shifting it from the middle leg to the corbicula on the hind leg on the same side of her body. Amazingly, while a worker bee is moving one resin lump to the corbicula of one hind leg—which requires moving three legs simultaneously and precisely—she is already detaching the next resin chunk or strand, which she will soon pack onto her other hind leg. Getting both hind legs loaded fully takes 15–60 minutes. The warmer the day, the faster the loading. I admire greatly the dexterity of worker bees "legging" (not handling!) lumps of sticky propolis, but of course, skillfully moving six legs simultaneously comes naturally to them, just as moving two legs and two arms simultaneously does to us.

Another aspect of resin collection that had been described in detail already is how a resin collector behaves when she gets home. In August 1926, Dr. Gustav A. Rösch—the second Ph.D. student of Karl von Frisch—watched the behavior of four worker bees that were members of a colony living in an observation hive and were visiting an artificial resin source that he had set up in the garden courtyard of the Zoological Institute in Munich (a bit of which is shown in Figure 16.1). Rösch had labeled these four bees for individual identification, so he was able to track them closely in his hive. He reported that when a resin collector arrives home, she does not join the crowd of bees on the combs just inside the hive entrance (the "dance floor"), in the manner of a nectar collector. Instead, a *resin collector* walks to a place where resin is being used to caulk cracks and fill crevices, and then she stands there and waits for other workers (*resin users*) to come to her and remove her resin load, bit by bit. Once a resin user has removed a small piece from a resin collector's load, she walks to a work site and uses her mandibles to press her bit of resin into her chosen spot. Then this resin-user bee, or another bee, nibbles wax particles off nearby combs and works these particles into the resin, to create a resin-wax mixture—propolis. It is less sticky and more rigid than pure resin.

Rösch reported that the length of time that a resin collector waits to get fully unloaded varies greatly, from one to seven hours, and he suggested that a resin collector's unloading time informs her of her colony's need for more resin. This "unloading difficulty hypothesis" proposes an *indirect mechanism* of sensing the need for resin. If it is correct, then nectar collectors, water collectors, and resin collectors all use the same mechanism to sense their colony's need for their material (as discussed for nectar and water in Chapters 14 and 19). A high need is indicated by a quick unloading. It should be noted, though, that the durations of the unloading delays differ: for nectar collectors and water collectors, they are measured in seconds, but for resin collectors they are measured in minutes or even hours.

Waltraud Meyer, however, suggested that resin collectors have a *direct mechanism* of sensing their colony's need for resin. In her Ph.D. studies in Berlin in the 1950s, she observed that worker bees that collect resin (outside the hive) often also work with resin (inside the hive). This happens when a resin collector comes home holding a gob of resin in her mandibles and then she uses it herself to coat a surface or fill a crevice. Meyer referred to this activity as "cementing," and in her studies she found that most of the bees doing this were forager-age bees, i.e., in the same age range as resin collectors (Fig. 20.3). This "cementing" work might give a resin collector a clear sense of her colony's need for resin. So, Jun and I wondered, who is right? Rösch? Meyer? Maybe both? We knew that a fresh look at the in-hive behaviors of the resin collectors *and* the resin users was needed. Who are these bees? And, most importantly, how common is it for a worker bee to function as *both* a resin collector and a resin user on the same day?

On 27 April 2002, Jun and I installed a colony of approximately 3,000 bees in a small (two-frame) observation hive that was housed in a warm room at my laboratory. This colony was the focus of Jun's research for the next six months, until the end of October that year.

Our first goal was to determine the ages at which worker bees perform the two aspects of resin work: (1) collecting it outside the hive and (2) using it inside the hive. This investigation began when we introduced eight 100-bee cohorts of 0-day-old workers to the observation-hive colony at 3–4 day intervals across the period of 5–27 May 2002. Each bee in each cohort was labeled for individual identification, by giving them various paint marks and color-number tags. Jun made observations of these bees every day from 8 May to 23 June. This involved scanning both sides of the hive for labeled bees that were performing any of the three resin-use activities: *detecting crevices* by inserting the antennae in them, *caulking crevices* by forcing resin into them with the mouthparts, and

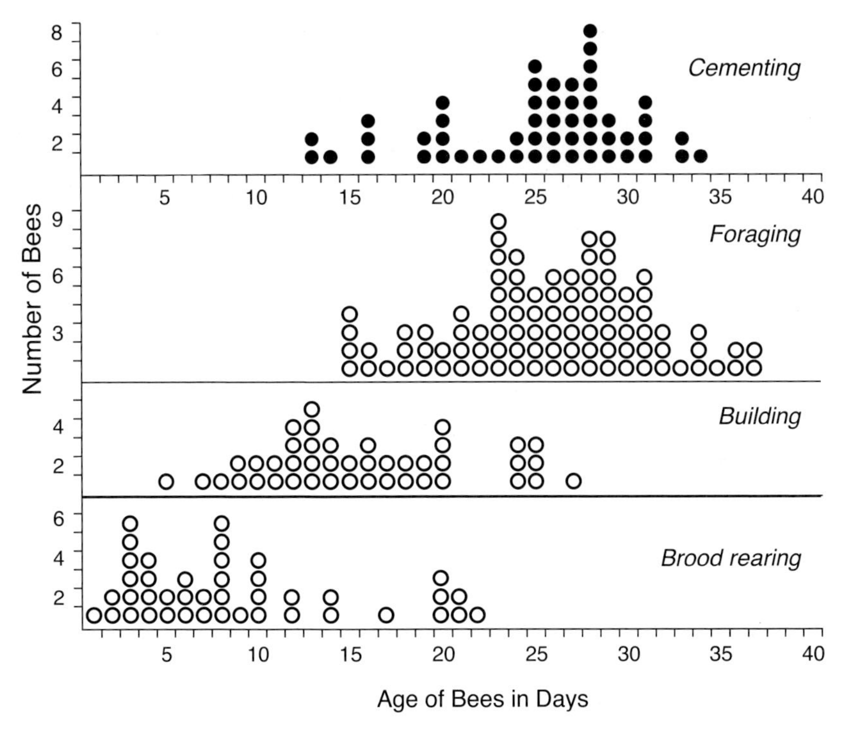

FIG. 20.3. The ages of resin-user ("cementing") bees, 27–29 August 1953. The ages of foraging bees, comb building bees, and brood rearing bees were also recorded, for comparison. The brood rearing bees were mostly young bees, comb building bees were mostly middle-age bees, and the foraging and the cementing bees were mostly elderly bees.

taking resin from a resin collector by biting some from one of her hind legs.

Marked bees were rarely seen performing resin-use tasks in May and June. Of the 800 individually identifiable bees that Jun introduced, only 10 were observed working with resin. They did so mostly when they were 14–24 days old, which was *after* they had stopped working as nurse bees (e.g., feeding larvae and eating pollen) and *before* they began working as foragers. (These 10 bees worked as foragers when they were 25–38 days old.) Also, none of these 10 resin-user bees became a resin-collector bee when she switched to working outside the hive. So, this first study showed us that some of a colony's resin-use work is performed by middle-age workers, but it did not show us the ages of the colony's resin collectors.

We supposed that they are mainly elderly bees, just like a colony's nectar collectors and pollen collectors.

Our next endeavor was to observe the in-hive behavior of bees that were resin collectors, for we needed to see how these bees behave when they get home carrying loads of resin. Would we see, as Waltraud Meyer had reported, that the division of labor between collectors and users is less clear-cut for bees doing resin work than for bees doing nectar work and water work? In other words, would we see that some resin collectors also function as resin users? Most of this work was done in September and October. With winter approaching, there was a good traffic of resin collectors dashing into the hive, bearing the shiny loads of resin their colony needed to seal up its home.

Jun devised an ingenious setup for labeling the resin collectors in our observation-hive colony (Fig. 20.4). It consisted of a traffic divider that was mounted outside the hive's entrance and that separated the exiting and entering bees. Bees exiting the hive were funneled to leave through a narrow opening atop a wall in the outer end of the traffic divider. Bees entering the hive were funneled to come inside through a wide opening in the outer end of the device. (It worked beautifully; the separation of exiting and entering bees was 99.9 percent effective.) The traffic divider had a clear top, so it was easy to spot the entering bees that were carrying resin loads . . . the resin collectors! A mirror helped Jun spot the resin collectors that were walking into the hive while upside down on the traffic divider's glass cover. Whenever Jun spotted an unmarked resin collector entering the hive, he deftly pinned her between a small net and a soft, plastic board, and then he marked her through the net with color paint pens. He labeled 102 resin collectors between 1 September and 6 October 2002.

Jun recorded the departure and arrival times of the individually identifiable resin collectors, observed their activities inside the hive, and recorded (on sheets of clear polyester film taped to the observation hive) such things as where these bees got unloaded, performed waggle dances, and received food. Jun also made video recordings of 35 individually

FIG. 20.4. Traffic divider outside the entrance to our observation hive. It separated the exiting and entering bees. This made it fairly easy to spot bees heading inside with loads of resin, and then to capture them (briefly) to label them with paint pens. The mirror helped Jun spot resin collectors that were entering the hive by walking upside down on the glass cover.

identifiable resin collectors starting when they entered the hive, to closely study their behavior inside the hive by means of slow-motion playback (Fig. 20.5).

Jun's painstaking work revealed something very telling about the organization of resin work: when a *resin* collector gets home, she walks deep inside the hive, to where the resin users are working (Fig. 20.6). This behavior is strikingly different from what a *nectar* collector does when she gets home, for usually she walks only a few inches (10–20 centimeters) inside her home before she is met by a nectar receiver. This difference in unloading locations is functional. Resin gets unloaded slowly, in stages, because each resin user removes just a snippet of a resin collector's two loads. So, by walking to a site where resin users are working, a resin collector minimizes the total distance that the resin users need to

FIG. 20.5. Professor Jun Nakamura video recording a resin collector in the observation hive. With this recording, he was able to make a detailed analysis of her behavior.

walk to get their resin bits. This helps minimize the total energy expended by resin users to process a resin collector's load. In contrast, a nectar collector gets unloaded quickly, and often in a single regurgitation to one nectar receiver. So, if a nectar collector were to walk up to the combs where nectar receivers are storing nectar, she would not save multiple nectar receivers from walking down to the nectar unloading area (near the hive entrance).

We see, too, in Figure 20.6, that the *resin* collectors and *nectar* collectors had spatially distinct "dance floors." Here, again, this difference is functional. This spatial segregation probably helps a bee that is starting work as a resin (or nectar) collector to get information about a good source of the material that she seeks. It is analogous to the way that Amazon's website is organized with different "places" for presenting information about hardware and groceries.

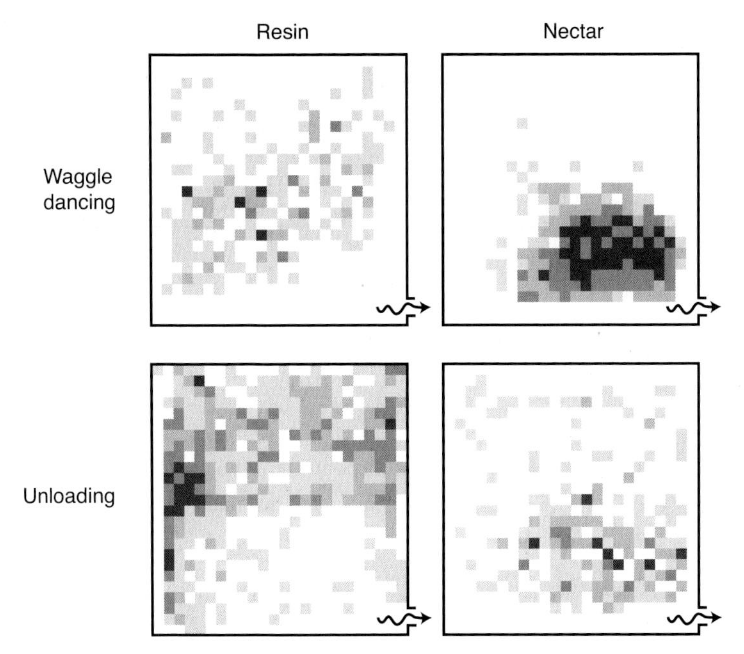

FIG. 20.6. Spatial distributions of the locations of waggle dancing and unloading for resin collectors and nectar collectors in the two-frame observation hive that we used for studies of resin collection. The wiggly arrow at the bottom right of each plot marks the hive's entrance. Within each plot, the darker the square, the greater the waggle dancing (upper drawings) or the unloading (lower drawings).

Further observations of the 102 individually identifiable resin collectors revealed more features of their behavior. One is that these bees have a high, but not absolute, fidelity to the task of resin collecting. Overall, 68 (67 percent) of the 102 bees that were labeled as resin collectors continued to collect resin for as long as they came and went from our hive, that is, for the remainders of their lives. The bee that performed this task the longest did so for 18 days. Bees that did not stick to resin collection usually switched to pollen collection.

By watching the 102 labeled bees come and go as resin collectors and, in some cases, eventually as pollen collectors, we were able to compare the outside-hive and inside-hive times per collecting trip for resin collectors and pollen foragers. This revealed that the collectors of these two materials spent the same amounts of time (average) outside the hive per collecting trip: *resin*, 35 minutes, based on 194 trips; *pollen,* 33 min, based on 119 trips. But they spent markedly different amounts of time (average) inside the hive per collecting trip: *resin*, 31 minutes, based on 151 trips; *pollen*, 11 minutes, based on 213 trips. Why is there this difference? A resin collector gets unloaded bit by bit, so often she spends much time standing around inside the hive between collecting trips. A pollen collector has an easier unloading experience: just find a suitable cell and "kick" the pollen loads off into it. For her, it is a speedy, drop-and-go operation.

It was fascinating to watch and learn about how resin collectors behave when they get home, but our primary purpose in tracking their in-hive behavior was to understand how a colony's resin collection is regulated. On the one hand, Gustav Rösch had suggested, in 1927, that a resin collector senses her colony's need for more resin indirectly, by noting how long it takes to get unloaded by resin users. On the other hand, Waltraud Meyer had suggested, in 1954, that a resin collector can sense her colony's need for resin directly, while "cementing," that is, while filling a crevice with resin that she has carried home. If she easily finds a place that needs resin work, then perhaps this tells her that her colony needs more resin. Jun and I wondered if *both* Rösch and Meyer might be right. In other words, we wondered whether resin collectors might be clever enough to pay attention to both the difficulty of getting unloaded and—if she carries home some resin gripped with her mandibles—the difficulty of finding a spot to caulk.

Happily, observing the in-hive activities of the 102 labeled resin collectors shed light on how resin collectors sense the need for their work. Every afternoon, for one to two hours, Jun watched inside the observation hive to see what the labeled resin collectors did between their collecting trips outside the hive. Of the 102 labeled resin collectors, Jun watched 77 throughout one or more returns to the hive. All had their resin loads removed from their hind legs by resin users. This was not surprising. After all, a resin collector, unlike a pollen collector, cannot simply kick off her sticky loads of resin. What was surprising, though, was seeing that many of the resin *collectors* also functioned as resin *users*: 16 (21%) were seen caulking, 7 (9%) were seen scraping wax with their mandibles to get wax to mix with resin for caulking, 6 (8%) were seen taking resin from a fellow resin collector, and 6 (8%) were seen examining crevices with their antennae (a key part of the caulking process). Seeing resin collectors function also as resin users confirmed Meyer's report that a high percentage—she saw 30% and we saw about 40%—of a colony's resin collectors also function as resin users. We concluded that resin collectors have many opportunities to sense their colony's need for propolis both indirectly and directly.

This conclusion is supported further by several things that we observed in the video recordings that Jun made of the in-hive behavior of 35 resin collectors. Each recording began when a labeled resin collector scurried into the hive. One curious thing we saw clearly is that a *resin* collector, unlike a *nectar* collector, does not remain in the thick crowd of bees on the combs just inside the hive's entrance (the "dance floor"). Instead, as is shown in Figure 20.7, which depicts the in-hive actions of the resin collector NV03, she walks quickly along the edges of the combs to a location in the top or side of the nest where resin is being used to caulk cracks and crevices. In our hive, a walk by a resin collector to a site of resin use took, on average, only about 30 seconds. If no one starts to unload a resin collector, then she may perform a tremble dance, probably to call for an unloader. The first unloading of the resin collector NV03 came at 10:38, seven minutes after she arrived home. Then, between 10:40 and 10:55 she made six more pauses to get unloaded, and one pause to do caulking, before she walked to the other side of the hive, at 10:58.

A second curious thing that we saw in our video recordings of 35 resin collectors was that 5 of them came home with both a chunk of resin in their mandibles and two shiny loads of resin on their corbiculae. Each of these five bees began caulking with the resin in her mouthparts as soon as she reached a spot where caulking was being performed. So, here, too, we saw several resin workers functioning as both a collector and a user. The story was only slightly different for the other 30 resin collectors that

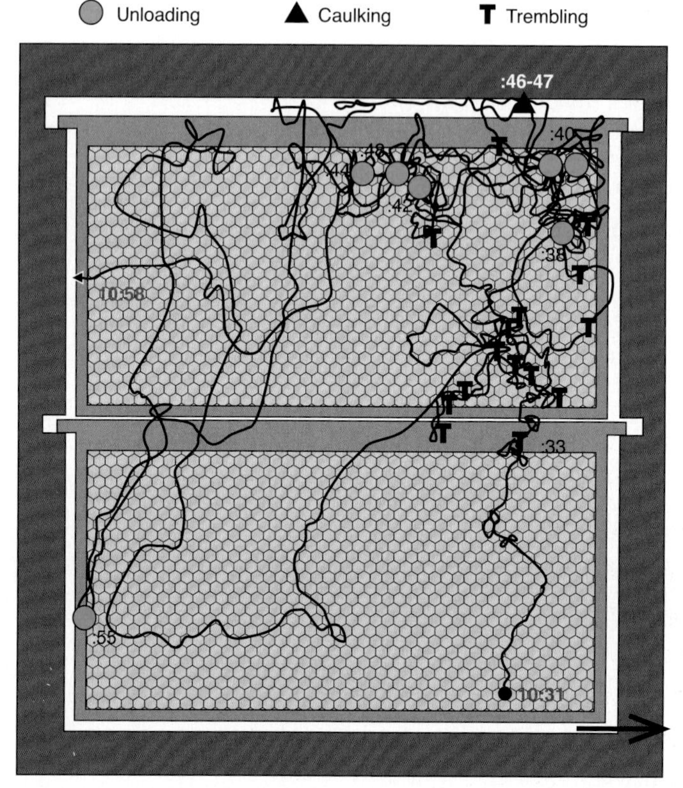

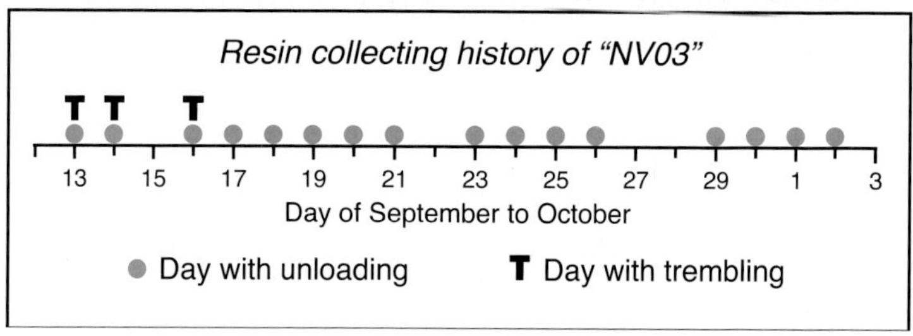

FIG. 20.7. *Top:* Pathway of resin collector NV03 during the first 27 minutes (from 10:31 a.m. to 10:58 a.m) of a return home on the second day of observing her work as a resin collector (September 14). Arrow at bottom right indicates the hive's entrance/exit. *Bottom:* Record of NV03's work as a resin collector on 17 days, between 13 September and 2 October.

brought their resin loads home entirely in their corbiculae. A sizable fraction (36 percent) of these bees engaged in caulking after other bees had plucked their resin loads, bit by bit, from their hind legs.

These observations show that resin collectors can, and probably do, get information about their colony's need for resin both directly (by sensing places that need to be caulked, as proposed by Meyer) and indirectly (by sensing how long it takes to get unloaded, as proposed by Rösch). The stage is now set for experimental tests of the caulking activity and the unloading difficulty hypotheses. Do the resin collectors pay attention to both potential indicators of their colony's need for more resin? If so, then is one primary and the other secondary? I think it might be possible to artificially remove (with forceps) the loads carried home by resin collectors, and thereby manipulate their unloading times. If this is doable, then it will be fascinating to see whether resin collectors whose loads are removed shortly after they come into their home—an observation hive—show greater motivation to collect more resin (and perform waggle dances) compared to resin collectors whose loads are plucked off after a delay.

If future studies show that resin collectors do indeed use both direct and indirect mechanisms to stay informed of the need for resin, then this will reveal something important about the cognitive abilities of worker honey bees. It will show that these bees can be versatile and sophisticated in their means of acquiring information about their colony's need for more resin. Some biologists have favored the view that the workers in social insect colonies operate with minimal knowledge. I suspect, however, that worker honey bees integrate information from multiple indicators of the state of their colony, because this helps them sense their colony's needs accurately. I hope we don't have to wait long for someone to elucidate further how a colony's enterprising resin collectors monitor their colony's need for the structurally important, and wonderfully healthful, substances that they bring home.

Closing Thoughts

> The honey-bee is as far above the general run of other insects as man is above all his fellow mammals. The complexity of the social life of bees, their powers of mutual communication, their diversity in skills and employment, their debates and decisions on policy, are so remarkable that they raise the question of the capacity of the bee for thinking.
>
> —*Vincent B. Wigglesworth, "Is the honey-bee conscious?," 1987*

In 1962, Karl von Frisch wrote "The brain of a bee is the size of a grass seed and is not made for thinking. The actions of bees are mainly governed by instinct." The point he was making is that he doubted that a worker honey bee has *reflective consciousness*. In other words, he doubted that a honey bee has what a human being has: the ability to reflect on the sensations, feelings, and thoughts that are passing through his or her mind. I feel certain, however, that Karl von Frisch believed that a worker bee has a high level of *perceptual consciousness*, because he knew that she can sense and learn the scents, colors, and shapes of flowers. And he knew that she can sense and learn the appearance of her home. It was, after all, his earliest studies of honey bees, conducted in the 1910s and 1920s, that showed that a worker bee has these cognitive abilities. Furthermore, he knew from his studies of the waggle dance, conducted in the 1940s and 1950s, that a worker bee can sense and memorize the direction and distance of a rich flower patch relative to her home, and then can share this stored information with her hive mates by performing a waggle dance. This is a truly stunning feat of behavior. I believe that if Karl von Frisch were alive today, he would be delighted to learn that we now know that when a honey bee collects nectar, she notes not just the *direction* and *distance* of her nectar source but also its *desirability*, and that she expresses information about all these things (the three "d's") in the waggle dance that she performs when she gets home.

I hope that you have enjoyed learning about this recently recognized feature of the waggle dance, plus all the other impressive behavioral talents of worker honey bees that are described in this book. In the late 1960s, when I was a novice beekeeper, the prevailing view—and the one that I held at first—was that honey bees have simple behavioral repertoires and that they function with simple rules of behavior. But in the early 1970s, after closely reading Martin Lindauer's little book, *Communication among Social Bees,* I began to appreciate the astonishing behavioral complexity of worker honey bees, especially when they function as nest-site scouts. Now, having spent five decades studying how these bees function in various roles—nectar forager, water collector, resin collector, and nest-site scout—I see clearly, and I respect deeply, the behavioral sophistication and cognitive abilities of these little beings.

If you would like to see for yourself many of the behaviors that I have described in this book, then I encourage you to set up an observation hive, install a colony, and watch your bees. I suggest that you start by watching pollen foragers, one at a time, when they run into your hive. The bright loads of pollen carried by these bees will make them easy to follow. You will see how a pollen forager diligently inspects a string of open cells until she finds one holding some pollen, then she backs into it carefully, and finally she "kicks" off her pollen loads. Three other sorts of bees that are easy to watch and that show impressive behavioral skills are (1) guard bees standing at the hive's entrance with their forelegs raised, ready for action; (2) forager bees walking over the combs and pausing occasionally to shake other bees to rouse them to greater activity; and (3) resin collectors scooting to sites in your hive where resin work is underway. Yet another sophisticated suite of behaviors is that of the workers who stroke the queen with their antennae to acquire queen substance pheromone, and then walk quickly around the hive to spread this signal of her presence. And if there is a nectar flow underway, then perhaps you will spy worker bees that seem to be suffering from some nervous system disorder, but are actually producing a fully functional signal: the tremble dance.

If it is not feasible for you to set up an observation hive, then I encourage you take advantage of a natural opportunity to observe closely the behavior of honey bees: a swarm that has settled in a spot where you can sit beside it and watch the bees' activities on the surface of its cluster. You will be able to see nest-site scouts performing their waggle dances, listen to them when they produce their piping signals in preparation for takeoff, and observe them perform their feverish buzz-runs just before the swarm launches into flight. A swarm that is "hanging out" where its members can be watched up close presents a marvelous opportunity to see firsthand the behavioral complexity of these bees.

Honey bees have captured the interests of human beings for millennia, and I expect that they will be a prime source of fascination for us forever. They are the makers of honey, and they are paragons of cooperation. I hope that you have enjoyed retracing with me twenty of my favorite trails through the scientific terrain of honey bee behavior.

Appendix: List of Signals

Names and functions of the signals of honey bees

(Those in italics are not discussed in this book)

Mechanical Signals	*Function*
Waggle dance	Recruit workers to locations (sources of nectar, pollen, water, and resin; and potential nest sites)
Tremble dance	Recruit workers to work as nectar receivers
Grooming invitation dance	Invite grooming by another worker
Shaking (DVAV)	General activation of workers
Antennation	Solicit food or attention from a worker
Worker piping (with wings together)	Stimulate other bees to warm their flight muscles
Buzz-run	Stimulate swarm departure from home/cluster
Streaker flight	Indicate direction of swarm's flight to its new home
Queen piping	Indicate queen's presence to workers
Worker beeping, aka "stop signal"	*Inhibit workers from performing waggle dances*
Worker piping (with wings apart)	*Function not known*

Chemical Signals	*Function*
Worker Nasonov gland pheromone	Attract nestmates to a location
Worker sting gland pheromone	Alarm nestmates
Queen mandibular gland pheromone	Indicate presence of queen
Worker footprint pheromone	*Indicate nest entrance*
Worker brood pheromone	*Indicate presence of brood*
Worker mandibular gland pheromone	*Elicit aggression*
Queen Dufour gland pheromone	*Mark queen-laid egg*
Queen tergal gland pheromone	*Attract workers?*
Queen footprint pheromone	*Inhibit queen rearing?*

Notes

CHAPTER 1. AVOIDING ASPHYXIA

Page 5: The report by Hazelhoff that is mentioned here is Hazelhoff (1941).

Page 6: See Ribbands (1953), page 212, for his summary of Hazelhoff's study.

Pages 6–7: See Lacher (1964) for the full report of his studies of the olfactory cells on the antennae of worker honey bees.

Pages 7–9: See Seeley (1974) for detailed information about my study of the carbon dioxide levels in two colonies.

Page 9: Information about the recommended limit of CO_2 in a workplace is found at https://www.fsis.usda.gov/sites/default/files/media_file/2020-08/Carbon-Dioxide.pdf 6.

Page 10: See Peters et al. (2017) and Peters et al. (2019) for detailed information about the differences in the bees' wing movements while fanning vs. while flying, and about the velocities of the air streams they produce for nest ventilation.

Page 13: See Hazelhoff (1954) for the posthumous report in a scientific journal of E. H. Hazelhoff's studies of nest ventilation by honey bees, published originally in a beekeeping magazine.

CHAPTER 2. FOREST HOMES

Page 14: See Crane (1978) and Kritsky (2010) for histories of the development of bee hives.

Pages 14–15: See Seeley (2019) for a detailed discussion of why it is incorrect to think of honey bees as domesticated animals.

Page 15: See Seeley (2017) and Seeley (2019) for the evidence that honey bee colonies still survive on their own in tree cavities and other nest sites, so not just in beekeepers' hives.

Page 15: See Hernández-Pacheco (1924) for information about the cave paintings in Spain that depict honey hunters in action thousands of years ago.

Page 15: The fascinating investigations of house hunting by honey bees, conducted by the German biologist Martin Lindauer, are reported in Lindauer (1955).

Pages 16–21: The investigations of the natural nests of honey bee colonies living in the woods around Ithaca, New York are reported in Seeley and Morse (1976).

Pages 19–20: For more information on what I have learned about how to be a bee-friendly beekeeper, see Seeley (2017a) and Seeley (2019).

Pages 21–28: The full report of the study of the nest-site preferences of honey bees is Seeley and Morse (1978). This investigation is also discussed in good detail in Chapter 5 in Seeley (2019).

CHAPTER 3. HOMESITE INSPECTORS

Page 28: The splendid book that is discussed here is Lindauer (1961).

Page 29: The biography of Lindauer is Seeley et al. (2002).

Page 30: The 62-page paper by Lindauer (translated title: "Swarm bees out house hunting") is Lindauer (1955).

Page 30: My Ph.D. thesis is Seeley (1978).

Page 33: The question of what nest-site scouts seek is addressed in Seeley and Morse (1978). What is known about how these bees sense the size of a potential nest cavity is reported in Seeley (1977).

Page 34: The distribution of nest-cavity volumes shown in Fig. 3.2 is from Seeley and Morse (1976).

Pages 34–40: All the studies conducted on Appledore Island in the summers of 1975 and 1976 are reported in detail in Seeley (1977).

Page 40: The "little book" by Karl von Frisch, mentioned here, is Frisch (1950). For a detailed discussion of the evidence for a worker honey bee being a "remarkably canny little creature," see Chittka (2022).

CHAPTER 4. CHOOSING A HOMESITE

Page 41: The figure of three-quarters of the bees in a colony departing in the swarm comes from the study by Rangel and Seeley (2012) of what fraction of a colony leaves when it casts a swarm. The figure of 12,000 bees, on average, in a swarm comes from Fell et al. (1977).

Page 41: The deadly duels between virgin queens are described in Gilley and Tarpy (2005).

Page 42: Latham (1927) is an example of a beekeeper who suspected that it is nest-site scouts that produce the conspicuous dances that are performed on swarms of honey bees.

Page 42: The marvelous investigations by Martin Lindauer are reported (in German) in Lindauer (1955). They are also described (in English) in Seeley (2010).

Pages 46–50: The studies of swarm decision making, with swarms in which each bee was labeled for individual identification, are described in Seeley and Buhrman (1999) and Seeley (2010).

Page 50: That a nest-site scout indicates the level of her enthusiasm for a potential nest site by the length of her advertisement with a waggle dance was shown by Seeley and Buhrman (2001).

Pages 50–52: The dance-off among nest-site scouts shown in Fig. 4.5 comes from Seeley (2010). This book discusses in detail how the nest-site scouts conduct their collective decision making.

CHAPTER 5. CONSENSUS OR QUORUM?

Page 54: The quote "There is no greater anomaly . . ." is from Chapter 5 in Darwin ([1859] 1964).

Pages 55–56: The description of Lindauer watching an airborne swarm splitting into two groups is based on what he reports in his marvelous paper Lindauer (1955).

Pages 56–60: The study in which Kirk Visscher and I watched an airborne swarm split into two groups of bees moving off in two opposite directions is reported in Seeley and Visscher (2003).

Pages 61–63: The study in which Kirk Visscher and I did an experimental test of the hypothesis that nest-site scouts rely on quorum sensing to know when they have reached a decision is reported in Seeley and Visscher (2004).

CHAPTER 6. PIPING HOT BEES

Page 68: The two prior studies of the temperatures inside swarm clusters are Büdel (1958) and Nagy and Stallone (1976).

Pages 68–69: The studies by Bernd Heinrich of thermoregulation in swarms are Heinrich (1981a) and Heinrich (1981b).

Pages 69–71: The study in which an infrared video camera was used to investigate how swarms warm up in preparation for takeoff is Seeley et al. (2003).

Page 72: The study of the expiration of dissent among the nest-site scouts on a honey bee swarm is Seeley (2003).

Pages 72–77: See Seeley and Tautz (2001) for detailed information about the piping signal that is used by nest-site scouts to stimulate the non-scouts in a swarm to warm up their flight muscles.

CHAPTER 7. BOISTEROUS BUZZ-RUNNERS

Page 79: See Butler (1609) for an early, but wonderful description of the exciting sight of a swarm of bees taking off and forming a cloud of airborne bees.

Page 79: The scientific paper by Martin Lindauer that is mentioned here is Lindauer (1955).

Page 81: See Esch (1967) for the earliest sound analysis of the buzz-run signal.

Page 81: The study by Clare Rittschof and colleagues of how a worker bee's developmental environment influences her future behavior is Rittschof et al. (2015).

Pages 81–83: See Rittschof and Seeley (2008) for the full report on the form, causes, and effects of the buzz-run signal.

Pages 82–83: See Bradbury and Vehrencamp (1998) for a clear discussion of the process of signal evolution (from incidental action to intentional signal) called ritualization.

CHAPTER 8. FLIGHT CONTROL IN SWARMS

Page 87: The pioneering studies of "queen substance" by Colin G. Butler are described in Butler (1954).

Pages 87–88: The study of the scent of 9-ODA as an indicator of the queen's presence in an airborne swarm is Avitabile et al. (1975).

Pages 88–91: The study of flight control in honey bee swarms that was conducted on Appledore Island in 1979 is Seeley et al. (1979).

Page 90: The study of the typical number of bees in a swarm is Fell et al. (1977).

Pages 91–96: The studies of swarm flights conducted in 2004—by Madeleine Beekman, Robert Fathke, Adrian Reich, and myself—are reported in Beekman et al. (2006).

Page 94: The 1975 paper mentioned here, with a speculation that the scouts in a swarm guide the other bees using the attraction pheromone, is Avitabile et al. (1975).

Pages 96–98: The study of the control of flight direction in swarms, which involved inducing a swarm to fly over a digital video camera and then using point-tracking software to extract (from the video recording) flight information about individual bees, is Schultz et al. (2008).

Page 97: The independent study of swarm flight control, conducted in Germany, is Greggers et al. (2013).

CHAPTER 9. ASTONISHING BEHAVIORAL VERSATILITY

Pages 101–102: For information about the form and function of worker piping signals, see Seeley and Tautz (2001) and Rangel and Seeley (2008).

Pages 101–103: See Rittschof and Seeley (2008) for our report on the form and effects of the buzz-run signal in a bivouacked swarm.

Pages 103–108: See Rangel et al. (2010) for the full report on the study that revealed the amazing behavioral sophistication of the nest-site scouts in a swarm.

CHAPTER 10. MESSENGER BEES

Page 110: For a revealing paper on the chemical basis of queen-worker communication, see Slessor et al. (1988).

Page 110: The role of 9-ODA as the sex attractant pheromone in honey bees was discovered in the early 1960s by Norman E. Gary, when a student at Cornell. See Gary (1962).

Page 110: See Carreck (2015) for an excellent memorial piece that describes Colin G. Butler's contributions.

Page 111: See Butler (1954) to see how it was determined that workers need to contact the queen to sense her presence. Airborne dispersal of her odor is not effective.

Page 111: See Gilley and Tarpy (2005) for a description and discussion of emergency queen rearing.

Page 112: See Allen (1960) for detailed information about the behavior of worker bees when they stand close to their queen (as her "attendants'), including that there is rapid turnover of these bees.

Page 113: See Verheijen-Voogd (1959) and Velthuis (1972) to read about the earliest evidence that workers will acquire some of a queen's pheromone and then will move around the colony's nest to disperse her signal.

Pages 113–114: My large observation hive, and the studies that I made with it to better understand how the queen's signal of her presence is dispersed widely and quickly among a colony's members, are described in Seeley (1979).

Pages 119–120: The virtuoso studies of the movement of 9-ODA in colonies, using radioactively labeled 9-ODA, are described in Naumann et al. (1991).

CHAPTER 11. A TALE OF FOUR SPECIES

Page 122: The biology, conservation issues, and cultural importance of the several species of honey bees that are native to Asia are described in good detail in Oldroyd and Wongsiri (2006).

Pages 122–123: See Grimaldi and Engel (2005) for more information on what the fossil record tells us about the origins and evolution of honey bees in the Asian tropics over the past 70 million years.

Page 124: See Chapter 4 in Darwin ([1859] 1964) to read about his comparative studies of blind cave beetles and sighted field beetles.

Pages 124–125: See Cullen (1957) to learn more about the beautiful comparative study of two closely related seabirds, black-legged kittiwakes and herring gulls.

Page 125: Karl von Frisch's autobiography is Frisch (1967b).

Page 126: The monograph in which we report the results of our comparative study of honey bees living wild in Thailand is Seeley et al. (1982).

Page 127: The investigations of the effects of worker-bee body size on the thermal biology of honey bees are reported in Dyer and Seeley (1987).

Pages 127–129: Michener (2000) provides a good, compact review of the evidence that the four species discussed here are closely related, and so are classified as members of the genus *Apis*.

Page 131: See Alexander (1991) and Engel and Schultz (1997) for the evidence that for honey bees, i.e., species in the genus *Apis*, nesting in the open is more ancient (ancestral) than nesting in cavities.

Pages 131–132: The information about the development times of the worker bees in the four species is found in Kapil (1959), Qayyum and Ahmad (1967), and Sandhu and Singh (1960).

Page 132: See Dyer and Seeley (1991) for the study of the relationship between nesting biology and worker lifespan in the different species of honey bees.

Page 134: The ability of honey bee colonies to defend themselves from large hornets by "cooking" them inside balls of worker bees is reported in Ono et al. (1987).

Page 135: The quote "most ferocious stinging insect on earth" is from Morse and Laigo (1969).

CHAPTER 12. COLONIES ARE INFORMATION CENTERS

Page 137: See Visscher (1983) to learn about how dead workers are recognized as such.

Page 137: See Heinrich (1979).

Page 138: See Frisch (1950) and Frisch (1967a).

Pages 138–142: The results of this study are reported in Visscher and Seeley (1982). Similar investigations of the long-range foraging abilities of honey bee colonies, made by spying on the waggle dances performed by foragers, have been conducted in England by Beekman et al. (2004) and Couvillon et al. (2014).

Page 143: The first study that I conducted at the Cranberry Lake Biological Station is described in Seeley (1986).

Pages 144–148: The test of a colony's skill in allocating its foragers wisely among nectar sources is Seeley et al. (1991).

Pages 148–149: The study that shows that an unemployed nectar forager follows a waggle dance chosen at random to learn where she should go to find work is Seeley and Towne (1992).

Page 149: The Honey Bee Algorithm is described in Nakrani and Tovey (2004).

CHAPTER 13. FORAGERS AS SENSORS

Page 150: See Eckert (1933), Visscher and Seeley (1982), and Beekman and Ratnieks (2000) for information about the flight range of foraging honey bees.

Pages 150–156: One study has explicitly investigated how the foragers in a colony function as its sensory units for attractive food sources. It is Seeley (1994).

Page 153: See Seeley and Towne (1992) for the evidence that when a forager follows a waggle dance to find a work site (a flower patch offering nectar or pollen), she follows a dance chosen at random among the dances performed in her colony's nest.

Page 154: The study that shows that honey bee colonies achieve a highly effective allocation of foragers among flower patches is Bartholdi et al. (1993).

Page 157: See Towne and Moscrip (2008) to learn about the remarkable ability of worker bees to memorize the sun's direction in relation to landmarks around their home across all times of day.

Page 158: The evidence that an unemployed forager does not conduct comparison "shopping" among the dances performed in their nest is reported in Seeley and Towne (1992).

Page 160: The study that was done "Back in 1990" is reported in Seeley and Towne (1992).

CHAPTER 14. NECTAR FLOW ON?

Page 166: The quote of Karl von Frisch is from page 18 of Frisch (1967a).

Page 166: See Gary (2015) for discussion of the behavior of robber bees.

Pages 166–168: See Lindauer (1948) and Lindauer (1961) for his description of how the nectar foragers in a colony are less likely to produce waggle dances when nectar is abundant, i.e., during a nectar flow.

Page 168: See Seeley and Tovey (1994) for an operations research analysis of why the average unloading delay of nectar foragers increases when the traffic of nectar foragers is high, i.e., when there is a nectar flow.

Pages 168–171: The experiment that is described here is reported in detail in Seeley (1986).

Pages 172–173: See Seeley (1989) to learn the details of the study that examined how the unloading experience of a nectar forager differs between times of low and high rates of nectar collection by her colony.

Page 174: See Lloyd (1983) for a good discussion of the distinction between signals and cues as pathways of information flow between individuals.

Page 175: See Seeley (1998) for a list of signals and cues used by honey bees, and see Bradbury and Vehrencamp (2011) for a broad discussion of the topic of signals vs. cues.

CHAPTER 15. MYSTERY OF THE TREMBLE DANCE

Page 176: The quote of Ernst Spitzner is my translation of his words in German, as found in Frisch (1965). See Spitzner (1788) for the original source.

Page 176: See Chapter 7, Application of the Dances to Other Objectives, in Frisch (1967a) for a review of how bees use the waggle dance to share information about various sorts of targets, not just flower patches.

Pages 176–177: See Frisch (1923) for his first detailed report on the dances of honey bees. The quote about the tremble dance is my translation of his words on page 90 of this book-length report.

Page 178: This figure is from Seeley (1992).

Page 178: The quote of von Frisch about the tremble dance is from page 283 in Frisch (1967a).

Page 179: The study of how nectar foragers become less choosy about the flower patches they will advertise with waggle dances when nectar is sparse is Seeley (1992).

Pages 179–184: See Seeley (1989) for my report of the work that Mary, Oliver, and I did in the summer of 1987 at the Cranberry Lake Biological Station.

Pages 185–187: See Seeley (1992) for my study of what causes worker bees to perform tremble dances.

Pages 187–188: The biography mentioned here is Seeley et al. (2002).

Page 188: The findings shown in Fig. 15.5 are reported in Seeley (1992).

CHAPTER 16. TWO RECRUITMENT DANCES, OR JUST ONE?

Page 190: See "Chapter 1. Historical" in Frisch (1967a) for a compact summary of how KvF began his studies of the recruitment dances of honey bees. See Frisch (1967b) for a fuller description of his initial studies of these dances.

Page 192: The quote "I could scarcely believe my eyes . . ." is from Frisch (1967b). See pages 72–73.

Page 193: The quote "It seems reasonable to see the two types of dance as different expressions of the bee language. . . ." is my translation from German of his words on page 566 in Frisch (1920).

Page 193: See Frisch (1923).

Page 194: See Frisch (1980) for a detailed description of Brunnwinkl, in eastern Austria.

Pages 194–195: The experiments described here, performed by Beutler and von Frisch, are reported in Frisch (1946) and Frisch (1948).

Page 197: The recent studies that have looked closely at dances advertising nearby food sources, and have reported brief waggle phases in them, are Kirchner et al. (1988), Jensen et al. (1997), and Gardner et al. (2008).

Page 198: The quote that bees that have followed round dances "swarm out in all directions and examine the surroundings of the hive" is from Frisch (1967a). See page 46.

Pages 198–201: The study made in 2011, described here, is Griffin et al. (2012).

CHAPTER 17. MOVERS AND SHAKERS

Page 202: See Tinbergen (1963).

Page 203: See Sherman (1977).

Page 205: To see how the shaking signal was described and discussed in the 1940s and early 1950s, see Haydak (1945), Taranov and Ivanova (1946), Schick (1953), Istomina-Tsvetkova (1953), and Milum (1955).

Page 205: To see the various names that have been given to the shaking signal, see Milum (1955), Allen (1956), Frisch (1967), Gahl (1975), Fletcher (1978), and Schneider (1986).

Page 206: M. Delia Allen's initial paper on workers preparing a queen for swarming by shaking her is Allen (1956).

Page 206: M. Delia Allen's further observations on workers shaking their queen are reported in Allen (1958) and Allen (1959).

Page 207: The studies of Eleonore Hammann are reported in Hammann (1957).

Page 207: The findings from the Ph.D. studies of Stanley S. Schneider are reported in Schneider (1986).

Page 208: The later findings of Stanley S. Schneider and colleagues are reported in Schneider (1987) and Schneider et al. (1986).

Page 208: The work of James C. Nieh on the activating effects of shaking signals on foragers that receive them is reported in Nieh (1998).

Page 208: The research project of Susanne Kühnholz that is mentioned here, on the control of water collection, is reported in Kühnholz and Seeley (1998).

Page 210: The surprising observation of a forager performing shaking signals after her first return from a sugar-water feeder that had been left empty for two days (due to bad weather), is described in the Introduction section of Seeley et al. (1998).

Pages 211–214: The experimental study of the shaking signal conducted at the Cranberry Lake Biological Station in 1994 is reported in Seeley et al. (1998).

Pages 215–216: See Klein et al. (2008) and Klein and Busby (2020) for detailed information on the differences in sleep rhythms of elderly (forager) bees and younger (hive worker) bees.

Pages 216–217: See Klein et al. (2010) and Klein et al. (2018) for detailed information about the impaired precision of waggle dances produced by sleep-deprived foragers, and about how their less precise dances are less attractive to dance followers.

CHAPTER 18. GROOM ME, PLEASE

Pages 218–219: See Danforth et al. (2019) to get a comprehensive view of the diverse tools and techniques that bees—all bees, not just honey bees!—have for grooming pollen and debris from their bodies.

Pages 220–223: The pollen packing behavior of worker honey bees is described and depicted beautifully in Casteel (1912) and Hodges (1974).

Pages 224–225: The quote is from Haydak (1945).

Page 225: See Božič and Valentinčič (1995).

Pages 226–228: See Land and Seeley (2004) for the analyses of the causes and effects of performing a grooming dance.

Page 227: The figures for the average durations of a waggle dance (ca. one minute) and of a tremble dance come from Seeley (1995) (see pp. 90–92) and from Seeley (1992).

Page 229: See Esch (1961) and Wenner (1962) for the results of measuring the frequency of body waggling for bees producing waggle dances.

CHAPTER 19. COLONY THIRST

Page 231: The data shown in Fig. 19.1 were originally reported in Chapter 7 in Seeley (1995).

Page 232: See Fig. 9.2 in Seeley (1995) for a graphical comparison of the nectar and water sectors of a honey bee colony's economy.

Page 233: See Lindauer (1961).

Page 233: The original description of gobbeting by worker bees is in Park (1925).

Page 234: Fig. 19.2 is redrawn from Lindauer (1954).

Page 236: The report of Lindauer is my translation of part of a sentence on page 24 of his paper Lindauer (1954).

Pages 236–240: The results of this first experiment, shown in Fig. 19.5, are reported more fully in Kühnholz and Seeley (1997).

Page 240: See Chittka (2022) for an up-to-date report on what is known about the cognitive abilities of bees, especially worker honey bees.

Pages 241–243: The further trials of the heat-stress experiment, conducted at the Cranberry Lake Biological Station, are reported in detail in Kühnholz and Seeley (1997).

Page 243: For a broad review of the control of food and water intake by insects, see Bernays and Simpson (1982).

Page 243: See Nicolson (2009) for a review of what is known about the physiological basis of the homeostasis of the hemolymph (blood) of worker bees, including its osmotic stability even though these bees often fill their crops with dilute nectar and water.

Pages 245–248: See Ostwald et al. (2016) for details on the activation of water collectors when a colony's home is threatened with overheating.

CHAPTER 20. RESIN WORK

Page 249: See Bastos et al. (2008), Bilikova et al. (2013), and Wilson et al. (2015) for examples of studies of the toxicity of propolis to the bacterium that causes the most serious honey bee disease, American Foulbrood.

Page 249: These percentages regarding the blend of resin and beeswax in propolis come from Crane (1990).

Page 250: The heavy use of propolis to coat the surfaces inside tree cavities that are occupied by honey bee colonies is shown in Fig. 5.4 in Seeley (2019).

Page 250: See Simone-Finstrom and Spivak (2010) and Borba et al. (2015) for rigorous experimental studies that show the health benefits to honey bees of living in hives whose interior wall surfaces are coated with propolis.

Page 251: See Nakamura and Seeley (2006) for the deepest analysis available of how a colony controls its collection of resins to make propolis.

Page 252: See Meyer (1951), Meyer (1954), and Meyer (1956) to learn about her pathbreaking studies of the resin-loading behavior and other behaviors of resin collectors in honey bee colonies.

Page 253: See Rösch (1927) for preliminary, but pioneering, observations on the behavior of resin collectors.

Page 263: See Bonabeau et al. (1997) and Camazine et al. (2001) to learn about the perspective that the workers in social insect colonies operate with very limited information about the state of their colony.

CLOSING THOUGHTS

Page 264: The quote "The brain of a bee is the size of a grass seed . . ." comes from Frisch (1962).

References

Alexander, B.A. 1991. Phylogenetic analysis of the genus *Apis* (Hymenoptera: Apidae). *Annals of the Entomological Society of America* 84: 137–149.

Allen, M.D. 1956. The behaviour of honey bees preparing to swarm. *Animal Behaviour* 4: 14–22.

Allen, M.D. 1958. Shaking of honeybee queens prior to flight. *Nature* 181: 68.

Allen, M.D. 1959. The occurrence and possible significance of the "shaking" of honeybee queens by the workers. *Animal Behaviour* 7: 66–69.

Allen, M.D. 1960. The honeybee queen and her attendants. *Animal Behaviour* 8: 201–208.

Avitabile, A., R.A. Morse, and R. Boch. 1975. Swarming honey bees guided by pheromones. *Annals of the Entomological Society of America* 68: 1069–1082.

Bartholdi, J.J., III, T.D. Seeley, C.A. Tovey, and J.H. Vande Vate. 1993. The pattern and effectiveness of forager allocation among flower patches by honey bee colonies. *Journal of Theoretical Biology* 160: 23–40.

Bastos, E.M.A.F, M. Simone, D.M. Jorge, A.E.E. Soares, and M. Spivak. 2008. *In vitro* study of the antimicrobial activity of Brazilian propolis against *Paenibacillus larvae*. *Journal of Invertebrate Pathology* 97: 273–281.

Beekman, M., and F.L.W. Ratnieks. 2000. Long range foraging in the honey bee. *Functional Ecology* 14: 490–496.

Beekman, M., R.L. Fathke, and T.D. Seeley. 2006. How does an informed minority of scouts guide a honey bee swarm as it flies to its new home? *Animal Behaviour* 71: 161–171.

Beekman, M., D.J.T. Sumpter, N. Seraphides, and F.L.W. Ratnieks. 2004. Comparing foraging behaviour of small and large honey-bee colonies by decoding waggle dances made by foragers. *Functional Ecology* 18: 829–835.

Bernays, E.A., and S.J. Simpson. 1982. Control of food intake. *Advances in Insect Physiology* 16: 59–118.

Bilikova, K., M. Popova, B. Trusheva, and V. Bankova. 2013. New anti-*Paenibacillus larvae* substances purified from propolis. *Apidologie* 44: 278–285.

Bonabeau, E., G. Theraulaz, J.-L. Deneubourg, S. Aron, and S. Camazine. 1997. Self-organization in social insects. *Trends in Ecology and Evolution* 12: 188–193.

Borba, R.S., K.K. Klyczek, K.L. Mogen, and M. Spivak. 2015. Seasonal benefits of a natural propolis envelope to honey bee immunity and colony health. *Journal of Experimental Biology* 218: 3689–3699.

Bradbury, J.W., and S.L. Vehrencamp. 1998. *Principles of Animal Communication.* Sinauer, Sunderland, Massachusetts.

Božič, J., and T. Valentinčič. 1995. Quantitative analysis of social grooming behavior of the honey bee *Apis mellifera carnica. Apidologie* 26: 141–147.

Büdel, A. 1958. Ein Beispiel der Temperaturverteilung in der Schwarmtraube. *Zeitschrift für Bienenforschung* 4: 63–66.

Butler, C. 1609. *The Feminine Monarchie: Or, A Treatise Concerning Bees and the Due Ordering of Them.* Joseph Barnes, Oxford.

Butler, C.G. 1954. The method and importance of the recognition by a colony of honeybees (*A. mellifera*) of the presence of its queen. *Transactions of the Royal Entomological Society (London)* 105: 11–29.

Camazine, S., J.-L. Deneubourg, N.R. Franks, J. Sneyd, G. Theraulaz, and E. Bonabeau. 2001. *Self-Organization in Biological Systems.* Princeton University Press, Princeton, New Jersey.

Carreck, N.L. 2015. Colin G Butler, MA, PhD, FRPS, FIBiol, OBE, FRS (1913–2016). *Bee World* 92: 129–131.

Casteel, D.B. 1912. The behavior of the honey bee in pollen collecting. *United States Bureau of Entomology Bulletin* no. 121. Government Printing Office, Washington, D.C. (See also https://www.amazon.com/Behavior-Honey-Bee-Pollen-Collection/dp/1530002605?asin=1530002605&revisionId=&format=4&depth=1).

Chittka, L. 2022. *The Mind of a Bee.* Princeton University Press, Princeton, New Jersey.

Couvillon, M.J., R. Schürch, and F.L.W. Ratnieks. 2014. Waggle dance distances as integrative indicators of seasonal foraging challenges. *PLoS ONE* 9: e93495.

Crane, E. 1978. *The World History of Beekeeping and Honey Production.* Routledge, New York.

Crane, E. 1990. *Bees and Beekeeping. Science, Practice, and World Resources.* Cornell University Press, Ithaca, New York.

Cullen, E. 1957. Adaptations in the kittiwake to cliff nesting. *Ibis* 99: 275–302.

Danforth, B.N., R.L. Minckley, and J.L. Neff. 2019. *The Solitary Bees. Biology, Evolution, and Conservation.* Princeton University Press, Princeton, New Jersey.

Darwin, C.R. (1859) 1964. *On the Origin of Species: A Facsimile of the First Edition,* Chapter 4. Harvard University Press, Cambridge, Massachusetts.

Dyer, F.C., and T.D. Seeley. 1987. Interspecific comparisons of endothermy in honeybees (*Apis*): deviations from the expected size-related patterns. *Journal of Experimental Biology* 127: 1–26.

Dyer, F.C., and T.D. Seeley. 1991. Nesting behavior and the evolution of worker tempo in four honey bee species. *Ecology* 72: 156–179.

Eckert, J.E. 1933. The flight range of the honey bee. *Journal of Agricultural Research* 47: 257–285.

Engel, M.S., and T.R. Schultz. 1997. Phylogeny and behavior in honey bees (Hymenoptera: Apidae). *Annals of the Entomological Society of America* 90: 43–53.

Esch, H. 1961. Über die Schallerzeugung beim Werbetanz der Honigbiene. *Zeitschrift für vergleichende Physiologie* 45: 1–11.

Esch, H. 1967. The sounds produced by swarming honey bees. *Zeitschrift für vergleichende Physiologie* 56: 408–411.

Fell, R.D., J.T. Ambrose, D.M. Burgett, D. De Jong, R.A. Morse, and T.D. Seeley. 1977. The seasonal cycle of swarming in honey bees. *Journal of Apicultural Research* 16: 170–173.

Fletcher, D.J.C. 1978. The influence of vibratory dances by worker honeybees on the activity of virgin queens. *Journal of Apicultural Research* 17: 3–13.

Frisch, K. von. 1914. Der Farbensinn und Formensinn der Bienen. *Zoologische Jahrbücher. Abteilung für allgemeine Zoologie und Physiologie der Tiere* 35: 1–188.

Frisch, K. von. 1920. Über die "Sprache" der Bienen, I. Mitteilung. *Münchener Medizinische Wochenschrift* 20: 566–569.

Frisch, K. von. 1923. Über die "Sprache" der Bienen, eine tierpsychologische Untersuchung. *Zoologische Jahrbücher. Abteilung für allgemeine Zoologie und Physiologie der Tiere* 40: 1–186.

Frisch, K. von. 1946. Die Tänze der Bienen. *Österreichische Zoologische Zeitschrift* 1: 1–48.

Frisch, K. von. 1948. The dances of the honey bee. *Bulletin of Animal Behaviour*. Special Number: 1–31.

Frisch, K. von. 1950. *Bees: Their Vision, Chemical Senses, and Language.* Cornell University Press, Ithaca, New York.

Frisch, K. von. 1962. Dialects in the language of the bees. *Scientific American* 207: 78–89.

Frisch, K. von. 1967a. *The Dance Language and Orientation of Bees.* Harvard University Press, Cambridge, Massachusetts.

Frisch, K. von. 1967b. *A Biologist Remembers.* Pergamon Press, Oxford.

Frisch, K. von. 1980. *Fünf Häuser am See* [*Five Houses by the Lake*]. Springer Verlag, Berlin.

Gahl, R.A. 1975. The shaking dance of honey bee workers: evidence for age discrimination. *Animal Behaviour* 23: 230–232.

Gardner, K.E., T.D. Seeley, and N.W. Calderone. 2008. Do honeybees have two discrete dances to advertise food sources? *Animal Behaviour* 75: 1291–1300.

Gary, N.E. 1962. Chemical mating attractants in the queen honey bee. *Science* 136: 773–774.

Gary, N.E. 2015. Activities and behavior of honey bees. Chapter 10 in *The Hive and the Honey Bee,* edited by J.E. Graham. Dadant and Sons, Hamilton, Illinois.

Gilley, D.C., and D.R. Tarpy. 2005. Three mechanisms of queen elimination in swarming honey bee colonies. *Apidologie* 36: 461–474.

Greggers, U., C. Schöning, J. Degen, and R. Menzel. 2013. Scouts behave as streakers in honeybee swarms. *Naturwissenschaften* 100: 805–809.

Griffin, S.R., M.L. Smith, and T.D. Seeley. 2012. Do honeybees use the directional information in round dances to find nearby food sources? *Animal Behaviour* 83: 1319–1324.

Grimaldi, D., and M.S. Engel. 2005. *Evolution of the Insects.* Cambridge University Press, Cambridge, England.

Hammann, E. 1957. Wer hat die Initiative bei den Ausflügen der Jungkönigin, die Königin oder die Arbeitsbienen? *Insectes Sociaux* 4: 91–106.

Haydak, M.H. 1945. The language of the honeybees. *American Bee Journal* 85: 316–317.

Hazelhoff, E.H. 1941. De luchtverversching van een bijenkast gedurende den zomer. *Maandscrift voor Bijenteelt* 44: 10–14, 27–30, 45–48, 65–68.

Hazelhoff, E.H. 1954. Ventilation in a bee-hive during summer. *Physiologia Comparata et Oecologia* 3: 343–364.

Heinrich, B. 1979. *Bumblebee Economics*. Harvard University Press, Cambridge, Massachusetts.

Heinrich, B. 1981a. The mechanisms and energetics of honeybee swarm temperature regulation. *Journal of Experimental Biology* 91: 25–55.

Heinrich, B. 1981b. Energetics of honeybee swarm thermoregulation. *Science* 212: 565–566.

Hernández-Pacheco, E. 1924. *Las Pinturas Prehistóricas de Las Cuevas de la Araña (Valencia)*. Museo National de Ciencas Naturales, Madrid.

Hodges, D.H. 1974. *The Pollen Loads of the Honeybee*. Bee Research Association.

Istomina-Tsvetkova, K.P. 1953. [New data on the behavior of bees]. *Pchelovodstvo* 9: 15–23.

Jensen, I.L., A. Michelsen, and M. Lindauer. 1997. On the directional indications in the round dances of honeybees. *Naturwissenschaften* 84: 452–454.

Kapil, R.P. 1959. Variation in development period of the Indian bee. *Indian Bee Journal* 21: 3–6.

Kirchner, W.H., M. Lindauer, and A. Michelsen. 1988. Honeybee dance communication: acoustical indication of direction in round dances. *Naturwissenschaften* 75: 629–630.

Klein, B.A., and M.K. Busby. 2020. Slumber in a cell: honeycomb used by honey bees for food, brood, heating . . . and sleeping. *PeerJ* 8: e9583.

Klein, B.A., M. Vogt, K. Unrein, and D.M. Reineke. 2018. Followers of honey bee waggle dancers change their behaviour when dancers are sleep-restricted or perform imprecise dances. *Animal Behaviour* 146: 71–77.

Klein, B.A., A. Klein, M.K. Wray, U.G. Mueller, and T.D. Seeley. 2010. Sleep deprivation impairs precision of waggle dance signaling in honey bees. *Proceedings of the National Academy of Sciences, USA* 107: 22705–22709.

Klein, B.A., K.M. Olzsowy, A. Klein, K.M. Saunders, and T.D. Seeley. 2008. Caste-dependent sleep of worker honey bees. *Journal of Experimental Biology* 211: 3028–3040.

Kritsky, G. 2010. *The Quest for the Perfect Hive*. Oxford University Press, Oxford.

Kühnholz, S. and T.D. Seeley. 1998. The control of water collection in honey bee colonies. *Behavioral Ecology and Sociobiology* 41: 407–422.

Lacher, V. 1964. Elektrophysiologische Untersuchungen an einzelnen Rezeptoren für Geruch, Kohlendioxyd, Luftfeuchtigkeit und Temperatur auf den Antennen der Arbeitsbiene und der Drohne (*Apis mellifica* L.). *Zeitschrift für vergleichende Physiologie*. 48: 587–623.

Land, B.B., and T.D Seeley. 2004. The grooming invitation dance of the honey bee. *Ethology* 110: 1–10.

Latham, A. 1927. The mysteries of swarming. *Gleanings in Bee Culture* 55: 441–442.

Lindauer, M. 1948. Über die Einwirkung von Duft- und Geschmacksstoffen sowie anderer Faktoren auf die Tänze der Bienen. *Zeitschrift für vergleichende Physiologie* 31: 348–412.

Lindauer, M. 1954. Temperaturregulierung und Wasserhaushalt im Bienenstaat. *Zeitschrift für vergleichende Physiologie* 36: 391–432.

Lindauer, M. 1955. Schwarmbienen auf Wohnungssuche. *Zeitschrift für vergleichende Physiologie* 37: 263–324.

Lindauer, M. 1961. *Communication among Social Bees.* Harvard University Press, Cambridge, Massachusetts.

Lloyd, J.E. 1983. Bioluminescence and communication in insects. *Annual Review of Entomology* 28: 131–160.

Meyer, W. 1951. Über die Bauarbeiten an den Brut- und Honigzellen im Bienenvolk (*Apis mellifica* L.). Ph.D. Thesis. Freie Universität, Berlin.

Meyer, W. 1954. Die "Kittharzbienen" und ihre Tätigkeiten. *Zeitschrift für Bienenforschung* 2: 147–157.

Meyer, W. 1956. "Propolis bees" and their activities. *Bee World* 37: 25–36.

Michener, C.D. 2000. *The Bees of the World.* The Johns Hopkins University Press, Baltimore, Maryland.

Milum, V.G. 1955. Honey bee communication. *American Bee Journal* 95: 97–104.

Morse, R.A., and F.M. Laigo. 1969. Apis dorsata *in the Philippines.* Philippine Association of Entomologists, Laguna.

Nagy, K.A., and J.N. Stallone. 1976. Temperature maintenance and CO_2 concentration in a swarm cluster of honeybees, *Apis mellifera. Comparative Biochemistry and Physiology* 55A: 169–171.

Nakamura, J., and T.D. Seeley. 2006. The functional organization of resin work in honey bee colonies. *Behavioral Ecology and Sociobiology* 60: 339–349.

Nakrani, S., and C. Tovey. 2004. On honey bees and dynamic server allocation in internet hosting centers. *Adaptive Behavior* 12: 223–240.

Naumann, K., M.L. Winston, K.N. Slessor, G.D. Prestwich, and F.X. Webster. 1991. Production and transmission of honey bee queen (*Apis mellifera* L.) mandibular gland pheromone. *Behavioral Ecology and Sociobiology* 29: 321–332.

Nicolson, S.W. 2009. Water homeostasis in bees, with the emphasis on sociality. *The Journal of Experimental Biology* 212: 429–434.

Nieh, J.C. 1998. The honey bee shaking signal: function and design of a modulatory communication signal. *Behavioral Ecology and Sociobiology* 42: 23–36.

Oldroyd, B.P., and S. Wongsiri. 2006. *Asian Honey Bees: Biology, Conservation, and Human Interactions.* Harvard University Press, Cambridge, Massachusetts.

Ono, M., I. Okada, and M. Sasaki. 1987. Heat production by balling in the Japanese honeybee *Apis cerana japonica* as a defensive behavior against the hornet, *Vespa simillima xanthoptera* (Hymenoptera: Vespidae). *Experientia* 43: 1031–1032.

Ostwald, M.M., M.L. Smith, and T.D. Seeley. 2016. The behavioral regulation of thirst, water collection, and water storage in honey bee colonies. *The Journal of Experimental Biology* 219: 2156–2165.

Park, O.W. 1925. The storing and ripening of honey by honeybees. *Journal of Economic Entomology* 18: 405–410.

Passino, K.M., and T.D. Seeley. 2006. Modeling and analysis of nest-site selection by honeybee swarms: the speed and accuracy trade-off. *Behavioral Ecology and Sociobiology* 59: 427–442.

Peters, J.M., N. Gravnish, and S.A. Combes. 2017. Wings as impellers: honey bees co-opt flight system to induce nest ventilation and disperse pheromones. *Journal of Experimental Biology* 220: 2203–2209.

Peters, J.M., O. Peleg, and L. Mahadevan. 2019. Collective ventilation in honeybee nests. *Journal of the Royal Society Interface* 16: 20180561.

Phillips, M.G. 1956. *The Makers of Honey*. Crowell, New York.

Qayyum, H.A., and N. Ahmad. 1967. Biology of *Apis dorsata* F. *Pakistan Journal of Science* 19: 109–113.

Rangel, J., and T.D. Seeley. 2008. The signals initiating the mass exodus of a honey bee swarm from its nest. *Animal Behaviour* 76: 1943–1952.

Rangel, J., and T.D. Seeley. 2012. Colony fissioning in honey bees: size and significance of the swarm fraction. *Insectes Sociaux* 59: 453–462.

Rangel, J., S.R. Griffin, and T.D. Seeley. 2010. An oligarchy of nest-site scouts triggers a honeybee swarm's departure from the hive. *Behavioral Ecology and Sociobiology* 66: 979–987.

Ribbands, C.R. 1953. *The Behaviour and Social Life of Honeybees*. Bee Research Association, London.

Rittschof, C.C., and T.D. Seeley. 2008. The buzz-run: how honeybees signal "Time to go!" *Animal Behaviour* 75: 189–197.

Rittschof, C.C., C.B. Combs, M. Frazier, C.M. Grozinger, and G.E. Robinson. 2015. Early-life experience affects honey bee aggression and resilience to immune challenge. *Scientific Reports* 5: 15572.

Rösch, G.A. 1927. Beobachtungen an Kittharz sammelnden Bienen (*Apis mellifica* L.). *Biologisches Zentralblatt* 47: 113–121.

Sandhu, A.S., and S. Singh. 1960. The biology and brood-rearing activities of the little honeybee (*Apis florea* Fabricius). *Indian Bee Journal* 22: 27–35.

Schick, W. 1953. Über die Wirkung von Giftstoffen auf die Tänze der Bienen. *Zeitschrift für vergleichende Physiologie* 35: 105–128.

Schneider, S.S. 1986. The vibration dance activity of successful foragers of the honeybee, *Apis mellifera* (Hymenoptera: Apidae). *Journal of the Kansas Entomological Society* 59: 699–705.

Schneider, S.S. 1987. The modulation of worker activity by the vibration dance of the honeybee, *Apis mellifera*. *Ethology* 74: 211–218.

Schneider, S.S., J.A. Stamps, and N.E. Gary. 1986. The vibration dance of the honey bee. I. Communication regulating foraging on two time scales. *Animal Behaviour* 34: 377–385.

Schultz, K., K.M. Passino, and T.D. Seeley. 2008. The mechanism of flight guidance in honeybee swarms: subtle guides or streaker bees? *Journal of Experimental Biology* 211: 3287–3295.

Seeley, T.D. 1974. Atmospheric carbon dioxide regulation in honey-bee (*Apis mellifera*) colonies. *Journal of Insect Physiology* 20: 2301–2305.

Seeley, T.D. 1977. Measurement of nest cavity volume by the honey bee (*Apis mellifera*). *Behavioral Ecology and Sociobiology* 11: 287–293.
Seeley, T.D. 1978. Nest-site selection by honey bees. Ph.D. Thesis. Harvard University, Cambridge, Massachusetts.
Seeley, T.D. 1979. Queen substance dispersal by messenger workers in honeybee colonies. *Behavioral Ecology and Sociobiology* 5: 391–415.
Seeley, T.D. 1982. How honeybees find a home. *Scientific American* 247 (Oct.): 158–168.
Seeley, T.D. 1986. Social foraging in honey bees: how colonies allocate foragers among patches of flowers. *Behavioral Ecology and Sociobiology* 19: 343–354.
Seeley, T.D. 1989. Social foraging in honey bees: how nectar foragers assess their colony's nutritional status. *Behavioral Ecology and Sociobiology* 24: 181–199.
Seeley, T.D. 1992. The tremble dance of the honey bee: message and meanings. *Behavioral Ecology and Sociobiology* 31: 375–383.
Seeley, T.D. 1994. Honey bee foragers as sensory units of their colonies. *Behavioral Ecology and Sociobiology* 34: 51–62.
Seeley, T.D. 1995. *The Wisdom of the Hive*. Harvard University Press, Cambridge, Massachusetts.
Seeley, T.D. 2003. Consensus building during nest-site selection in honey bee swarms: the expiration of dissent. *Behavioral Ecology and Sociobiology* 53: 417–424.
Seeley, T.D. 2010. *Honeybee Democracy*. Princeton University Press, Princeton, New Jersey.
Seeley, T.D. 2017. Life-history traits of honey bee colonies living in forests around Ithaca, NY, USA. *Apidologie* 48: 743–754.
Seeley, T.D. 2017. Darwinian beekeeping: an evolutionary approach to apiculture. *American Bee Journal* 157: 277–282.
Seeley, T.D. 2019. *The Lives of Bees*. Princeton University Press, Princeton, New Jersey.
Seeley, T.D., and S.C. Buhrman. 1999. Group decision making in swarms of honey bees. *Behavioral Ecology and Sociobiology* 45: 19–31.
Seeley, T.D., and S.C. Buhrman. 2001. Nest-site selection in swarms of honey bees: How well do swarms implement the "best-of-*N*" decision rule? *Behavioral Ecology and Sociobiology* 49: 416–427.
Seeley, T.D., and R.A. Morse. 1976. The nest of the honey bee (*Apis mellifera* L.). *Insectes Sociaux* 23: 495–512.
Seeley, T.D., and R.A. Morse. 1978. Nest site selection by the honey bee, *Apis mellifera*. *Insectes Sociaux* 25: 323–337.
Seeley, T.D., and J. Tautz. 2001. Worker piping in honey bee swarms and its role in preparing for liftoff. *Journal of Comparative Physiology A* 187: 667–676.
Seeley, T.D., and C.A. Tovey. 1994. Why search time to find a food-storer bee accurately indicates the relative rates of nectar collecting and nectar processing in honey bee colonies. *Animal Behaviour* 47: 311–316.
Seeley, T.D., and W.F. Towne. 1992. Tactics of dance choice in honey bees: do foragers compare dances? *Behavioral Ecology and Sociobiology* 30: 59–69.

Seeley, T.D., and P.K. Visscher. 2003. Choosing a home: how the scouts in a honey bee swarm perceive the completion of their group decision making. *Behavioral Ecology and Sociobiology* 54: 511–520.

Seeley, T.D., and P.K. Visscher. 2004. Quorum sensing during nest-site selection by honeybee swarms. *Behavioral Ecology and Sociobiology* 56: 594–601.

Seeley, T.D., S. Camazine, and J. Sneyd. 1991. Collective decision-making in honey bees: how colonies choose among nectar sources. *Behavioral Ecology and Sociobiology* 28: 277–290.

Seeley, T.D., R. Hadlock Seeley, and P. Akratanakul. 1982. Colony defense strategies of the honey bees in Thailand. *Ecological Monographs* 52: 43–63.

Seeley, T.D., S. Kühnholz, and R.H. Seeley. 2002. An early chapter in behavioral physiology and sociobiology: the science of Martin Lindauer. *Journal of Comparative Physiology A* 188: 439–453.

Seeley, T.D., S. Kühnholz, and A. Weidenmüller. 1996. The honey bee's tremble dance stimulates additional bees to function as nectar receivers. *Behavioral Ecology and Sociobiology* 39: 419–427.

Seeley, T.D., R.A. Morse, and P.K. Visscher. 1979. The natural history of the flight of honey bee swarms. *Psyche* 86: 103–113.

Seeley, T.D., A. Weidenmüller, and S. Kühnholz. 1998. The shaking signal of the honey bee informs workers to prepare for greater activity. *Ethology* 104:10–26.

Seeley, T.D., M. Kleinhenz, B. Bujok, and J. Tautz. 2003. Thorough warm-up before take-off in honey bee swarms. *Naturwissenschaften* 90: 256–260.

Sherman, P.W. 1977. Nepotism and the evolution of alarm calls. *Science* 197: 1246–1253.

Simone-Finstrom, M., and M. Spivak. 2010. Propolis and bee health: The natural history and significance of resin use by honey bees. *Apidologie* 41: 295–311.

Slessor, K.N., L.-A. Kaminski, G.G.S. King, J.H. Borden, and M.L. Winston. 1988. Semiochemical basis of the retinue response to queen honey bees. *Nature* 332: 354–356.

Snodgrass, R.E. 1956. *Anatomy of the Honey Bee.* Cornell University Press, Ithaca, New York.

Spitzner, M.J.E. 1788. *Ausführliche Beschreibung der Korbbienenzucht im sächsischen Churkreise, ihrer Dauer und ihres Nutzens, ohne künstliche Vermehrung nach den Gründen der Naturgeschichte und nach eigener langer Erfahrung.* Junius, Leipzig.

Taranov, G.F., and L.V. Ivanova. 1946. [Observations upon queen behavior in bee colonies]. *Pchelovodstvo* 2/3: 35–39.

Tinbergen, N. 1963. On aims and methods of Ethology. *Zeitschrift für Tierpsychologie* 20: 410–433.

Towne, W.F., and H. Moscrip. 2008. The connection between landscapes and the solar ephemeris in honeybees. *Journal of Experimental Biology* 211: 3729–3736.

Velthuis, H.H.W. 1972. Observations on the transmission of queen substances in the honey bee colony by the attendants of the queen. *Behaviour* 41: 105–128.

Verheijen-Voogd, C. 1959. How worker bees perceive the presence of their queen. *Zeitschrift für vergleichende Physiologie* 41: 527–582.

Visscher, P.K. 1983. The honey bee way of death: Necrophoric behaviour in *Apis mellifera* colonies. *Animal Behaviour* 31: 1070–1076.

Visscher, P.K., and T.D. Seeley. 1982. Foraging strategy of honeybee colonies in a temperate deciduous forest. *Ecology* 63: 1790–1801.

Wenner, A. 1962. Sound production during the waggle dance of the honeybee. *Animal Behaviour* 10: 79–95.

Wigglesworth, V.B. 1987. Is the honey-bee conscious? *Antenna: Bulletin of the Royal Entomological Society of London* 11: 130.

Wilson, M.B., D. Brinkman, M. Spivak, G. Gardner, and J.D. Cohen. 2015. Regional variation in composition and antimicrobial activity of US propolis against *Paenibacillus larvae* and *Ascophaera apis*. *Journal of Invertebrate Pathology* 124: 44–50.

Illustration Credits

Fig. 1.1. Photo by Thomas D. Seeley.
Fig. 1.2. Original drawing by Margaret C. Nelson.
Fig. 1.3. Original drawing by Margaret C. Nelson, based on Fig. 17 in Lacher (1964).
Fig. 1.4. Original drawing by Margaret C. Nelson, based on data summarized in Table 1 in Seeley (1974).
Fig. 1.5. Photo by Jacob M. Peters.
Fig. 1.6. Original drawing by Margaret C. Nelson, based on Fig. 1 in Seeley (1974).
Fig. 2.1. Photo by Thomas D. Seeley.
Fig. 2.2. Photo by Thomas D. Seeley.
Fig. 2.3. Photo by Thomas D. Seeley.
Fig. 2.4. Purchased by Roger A. Morse in 1972 in a market in Kenya. It hung in his office at the Dyce Lab. The artist is unknown. Photo by Thomas D. Seeley.
Fig. 2.5. Photo by Thomas D. Seeley.
Fig. 3.1. Photo by John G. Seeley.
Fig. 3.2. Original drawing by Margaret C. Nelson, based on Fig. 2 in Seeley and Morse (1976).
Fig. 3.3. Photo by Robin Hadlock Seeley.
Fig. 3.4. Photos by Thomas D. Seeley.
Fig. 3.5. *Top:* Original drawings by Margaret C. Nelson, based on Fig. 2 in Seeley (1982).
Fig. 3.5. *Bottom:* Photo by Thomas D. Seeley.
Fig. 4.1. Photo by Kenneth Lorenzen.
Fig. 4.2. Original drawing by Margaret C. Nelson, based on Fig. 3 in Lindauer (1955).
Fig. 4.3. Photo by Thomas D. Seeley.
Fig. 4.4. Original drawing by Margaret C. Nelson, based on Fig. 1 in Seeley and Buhrman (1999).
Fig. 4.5. Original drawing by Margaret C. Nelson, based on Fig. 5 in Seeley and Buhrman (1999).
Fig. 5.1. Original drawing by Margaret C. Nelson, based on Fig. 7 in Lindauer (1955).
Fig. 5.2. Photo by Thomas D. Seeley.
Fig. 5.3. Original drawing by Margaret C. Nelson.
Fig. 5.4. Photo by Thomas D. Seeley.
Fig. 6.1. Photo by Thomas D. Seeley.

Fig. 6.2. Photo by Thomas D. Seeley.
Fig. 6.3. Photos by Thomas D. Seeley.
Fig. 6.4. Original drawing by Margaret C. Nelson, based on Fig. 2 in Seeley and Tautz (2001).
Fig. 6.5. Original drawing by Margaret C. Nelson, based on Fig. 8 in Seeley and Tautz (2001).
Fig. 6.6. Original drawing by Margaret C. Nelson, based on Fig. 7 in Seeley and Tautz (2001).
Fig. 7.1. Photo courtesy of Peter Essick.
Fig. 7.2. Original drawing by Margaret C. Nelson, based on Fig. 1 in Rittschof and Seeley (2008).
Fig. 7.3. Original drawing by Margaret C. Nelson, based on Fig. 2 and Fig. 3 in Rittschof and Seeley (2008).
Fig. 8.1. *Top:* Photo courtesy of James G. Morin.
Fig. 8.1. *Bottom:* Photo by Thomas D. Seeley.
Fig. 8.2. Photo by Thomas D. Seeley.
Fig. 8.3. Original drawing by Margaret C. Nelson, based on Fig. 2 in Beekman et al. (2006).
Fig. 8.4. Photo by Kenneth Lorenzen.
Fig. 8.5. Original drawing by Margaret C. Nelson, based on Fig. 4 in Schultz et al. (2008).
Fig. 9.1. Original drawing by Margaret C. Nelson, based on Fig. 1 in Rangel and Seeley (2008).
Fig. 9.2. Original drawing by Margaret C. Nelson, based on Fig. 4 in Rittschof and Seeley (2008)
Fig. 9.3. Photo by Thomas D. Seeley.
Fig. 9.4. Photo by Thomas D. Seeley.
Fig. 9.5. Original drawing by Margaret C. Nelson, based on Fig. 1 in Rangel et al. (2010).
Fig. 10.1. Original drawing by Margaret C. Nelson, based on Fig. 10 in Snodgrass (1956).
Fig. 10.2. Photo by Kenneth Lorenzen.
Fig. 10.3. Original drawing by Margaret C. Nelson, based on Fig. 1 in Seeley (1979).
Fig. 10.4. Original drawing by Margaret C. Nelson, based on Fig. 7 in Seeley (1979).
Fig. 10.5. Original drawing by Margaret C. Nelson, based on Fig. 8 in Seeley (1979).
Fig. 11.1. Original drawings by Sandra Olenik.
Fig. 11.2. Photo by Thomas D. Seeley.
Fig. 11.3 *Top/left:* Photo by Thomas D. Seeley.
Fig. 11.3 *Bottom/left:* Photo by Thomas D. Seeley.
Fig. 11.3 *Top/right:* Photo by Thomas D. Seeley.
Fig. 11.3 *Bottom/right:* Photo by Thomas D. Seeley.
Fig. 11.4. Photo by Thomas D. Seeley.
Fig. 12.1 Original drawing by Margaret C. Nelson, based on Fig. 3.1 in Seeley (1995).
Fig. 12.2. Photo by Thomas D. Seeley.
Fig. 12.3 Original drawing by Margaret C. Nelson, based on Fig. 3 in Visscher and Seeley (1982)

Fig. 12.4. Photo by Thomas D. Seeley.
Fig. 12.5. *Left:* Photo by Thomas D. Seeley.
Fig. 12.5. *Right:* Original drawing by Margaret C. Nelson, based on Fig. 5.26 in Seeley (1995).
Fig. 12.6. Original drawing by Margaret C. Nelson, based on Fig. 1 in Seeley et al. (1991).
Fig. 13.1. Photo by Thomas D. Seeley.
Fig. 13.2. Original drawing by Margaret C. Nelson, based on Fig. 5 in Seeley (1994).
Fig. 13.3. Original drawing by Margaret C. Nelson, based on Fig. 1 in Seeley (1994).
Fig. 13.4. *Left:* Photo by Thomas D. Seeley.
Fig. 13.4. *Right:* Original drawing by Margaret C. Nelson.
Fig. 13.5. Original drawing by Margaret C. Nelson, based on Fig. 3 in Seeley (1994).
Fig. 14.1 Original drawing by Margaret C. Nelson, based on Fig. 2.15 in Seeley (1995).
Fig. 14.2. Photo by Kenneth Lorenzen.
Fig. 14.3. *Left:* Original drawing by Margaret C. Nelson, based on Fig. 5.26 in Seeley (1995).
Fig. 14.3. *Right:* Photo by Thomas D. Seeley.
Fig. 14.4. Original drawing by Margaret C. Nelson, based on data in Table 1 in Seeley (1986).
Fig. 14.5. Original drawing by Margaret C. Nelson, based on data in Fig. 3 in Seeley (1989).
Fig. 15.1. Original drawing by Margaret C. Nelson, based on Fig. 1 in Seeley (1992).
Fig. 15.2. Original drawing by Margaret C. Nelson, based on Fig. 1 in Seeley (1989).
Fig. 15.3. Photo by Thomas D. Seeley.
Fig. 15.4. Original drawing by Margaret C. Nelson, based on Fig. 6 in Seeley (1992).
Fig. 15.5. Original drawing by Margaret C. Nelson, based on Fig. 7 in Seeley (1992).
Fig. 16.1. *Top:* Photo courtesy of the Bayerische Staatsbibliothek München/Bildarchiv.
Fig. 16.1. *Bottom.* Original drawing by Margaret C. Nelson, based on Plate 1, Fig. 4 in Frisch (1914).
Fig. 16.2. Original drawing by Margaret C. Nelson, based on Fig. 42 in Frisch (1950).
Fig. 16.3. Original drawing by Margaret C. Nelson.
Fig. 16.4. Original drawing by Margaret C. Nelson.
Fig. 16.5. Photo by Thomas D. Seeley
Fig. 17.1. Photo by Thomas D. Seeley
Fig. 17.2. Original drawing by Margaret C. Nelson.
Fig. 17.3. Original drawing by Margaret C. Nelson, based on Fig. 1 in Allen (1958).
Fig. 17.4. Original drawing by Margaret C. Nelson, based on a drawing by Barrett A. Klein.
Fig. 17.5. Original drawing by Margaret C. Nelson, based on data in Fig. 6 in Seeley et al. (1998).
Fig. 17.6. Photo by Barrett A. Klein.
Fig. 18.1. Original drawings by Margaret C. Nelson.
Fig. 18.2. Photo by Alex Wild.
Fig. 18.3. Original drawings by Margaret C. Nelson.

Fig. 18.4. Photo by Thomas D. Seeley.
Fig. 18.5. Original drawing by Margaret C. Nelson, based on Fig. 5 in Land and Seeley (2004).
Fig. 19.1. Original drawing by Margaret C. Nelson, based on Fig. 2.12 in Seeley (1995).
Fig. 19.2. Original drawing by Margaret C. Nelson, based on Fig. 9 in Lindauer (1954)
Fig. 19.3. Original drawing by Barrett A. Klein, based on a photo supplied by Fred C. Dyer.
Fig. 19.4. Photo by Thomas D. Seeley.
Fig. 19.5. Original drawing by Margaret C. Nelson, based on Fig. 2 in Kühnholz and Seeley (1997).
Fig. 19.6. Photo by Linton Chilcott.
Fig. 19.7. Photo by Thomas D. Seeley.
Fig. 20.1. Photo by Thomas D. Seeley.
Fig. 20.2. Photo by Kenneth Lorenzen.
Fig. 20.3. Original drawing by Margaret C. Nelson, based on Fig. 4 in Meyer (1954).
Fig. 20.4. Photo by Jun Nakamura.
Fig. 20.5. Photo by Keiko Nakamura.
Fig. 20.6. Original drawing by Margaret C. Nelson, based on Fig. 5 in Nakamura & Seeley (2006).
Fig. 20.7. Original drawing by Margaret C. Nelson, based on Fig. 6 in Nakamura and Seeley (2006)

Index